Science Fiction and Innovation Design

Innovation in Engineering and Technology Set

coordinated by
Dimitri Uzunidis

Volume 6

Science Fiction and Innovation Design

Edited by

Thomas Michaud

iSTE

WILEY

First published 2020 in Great Britain and the United States by ISTE Ltd and John Wiley & Sons, Inc.

ISTE Ltd
27-37 St George's Road
London SW19 4EU
UK

www.iste.co.uk

John Wiley & Sons, Inc.
111 River Street
Hoboken, NJ 07030
USA

www.wiley.com

Library of Congress Control Number: 2020938714

British Library Cataloguing-in-Publication Data
A CIP record for this book is available from the British Library
ISBN 978-1-78630-583-1

Contents

Introduction

Science Fiction: A Technical Imaginary World to be Deciphered

Innovative organizations evolve in ever-changing ideological and imaginary contexts. Understanding these flows is important in order to build an effective strategic vision. Economic actors must thus develop managerial discourse in relation to innovative concepts, some of which appear in popular culture, scientific imagination and science fiction. This artistic genre, increasingly diffused and appreciated in industrialized societies, is often presented as an inexhaustible source of concepts and prototypes of futuristic technologies. Futurology and foresight are taken into account in this imaginary world in order to better master and control the most promising sectors. Investing in a project, an organization or a company means adhering to a strategic vision. However, some of them are strongly inspired by science fiction aesthetics.

Science fiction appeared in the 19th Century with *Frankenstein* (1818) by Mary Shelley. Historians of the genre refer to earlier works, some of which date back to antiquity. However, it was the appearance of utopia, a term coined by Thomas More in 1516, that marked a break in the way of conceiving the imaginary world in the West. From then on, artists and intellectuals began to become aware of the impact of fiction on the way the world is not only understood but also constructed. Political power then became suspicious of potentially subversive narratives. *Micromégas* by Voltaire (1742) is thus presented as a work of prescience fiction philosophy. The 19th Century saw a great deal of activity in the creation of scientific

Introduction written by Thomas MICHAUD.

fiction. In France and Great Britain, a large number of stories were published, reflecting an intellectual ferment in the developing territories of the Industrial Revolution. Utopian thought, associated with industrialism and the works of authors such as Jules Verne, helped to influence political and scientific thought. Works of fiction were beginning to question the meaning of the Industrial Revolution. The increase of innovations provoked the expression of passions, both positive and negative. Since its beginnings, the imagination has accompanied a constantly changing industrial revolution.

The term science fiction (scientifiction) was coined in the United States. Hugo Gernsback contributed greatly to the development of this genre by publishing numerous journals in which a large number of utopian technologies appeared. Science fiction then spread rapidly, gradually finding its place in the dialogue between the humanities and the so-called hard sciences. It developed original approaches, questioning the relationship of decision-makers, whether political or economic, to a technical progress sometimes described as Promethean by some of its critics. Cinema consecrated the success of science fiction: special effects made it possible to represent improbable technologies or scientific phenomena, fascinating many people interested in science and technology. Note that 200 years after the publication of Mary Shelley's work, science fiction is experiencing a tremendous boom, which should enter a new era with the significant drop in the price of special effects. In the coming years and decades, the art of creating futuristic worlds in cinema should become affordable to a large number of actors, not only to a Hollywood elite.

Two approaches to science fiction emerge from this context. On the one hand, innovators are increasingly interested in how science fiction writers have imagined the future of their industries. Some believe in its anticipatory, futuristic, even prophetic dimension. A better knowledge of the history of utopian technologies is necessary for them to develop their strategic discourse and innovation policies. The relationship between science and science fiction should therefore be examined in order to determine how the interactions between two disciplines that are central to the innovation processes of technoscientific societies work. On the other hand, more and more economic actors are deciding to create their own science fiction, institutional science fiction, with the aim of no longer realizing the visions of external authors, but promoting utopian technologies imagined internally.

Science fiction is a particularly popular type of fantasy in the discourse of innovators and entrepreneurs in many sectors such as converging technologies or the space industry. Organizations must adapt to constantly evolving imaginations, utopian or dystopian, but also very influential in the definition of strategies and objectives for Research and Development (R&D). The science fiction imaginary world is particularly influential in the most innovative territories. The United States and Japan regularly offer works of fiction that fascinate a considerable number of individuals and minds passionate about technosciences. An imaginary pact unifies the visions of the future of an international community of engineers and scientists behind innovations that are radically transforming lifestyles.

How are engineers influenced by the imagination, and particularly by science fiction? Is scientific rationality compatible with this culture, since the imagination is sometimes denounced as a drift that is harmful to reason? How do organizations deal with science fiction, for example, in large telecommunications, nanotechnology and biotechnology groups, but also in aeronautics, aerospace and many other sectors? The number of cases of organizations, companies and institutions using science fiction for foresight purposes is growing every day. Which works play or have played an important role in the imagination of engineers and scientists? According to what psycho-sociological mechanisms?

Although capitalism regularly experiences crises, these crises testify to the failure of imaginary worlds that have, over a period of time, supported the development and dissemination of innovations, particularly technological ones. Science fiction can be seen as an imaginary world derived from the archetypes at the origin of scientific thought. Is this position – inspired by the theories of the imagination of thinkers such as Gaston Bachelard, Gilbert Durand or Cornelius Castoriadis, but also by the theory of archetypes and the collective unconscious of C.G. Jung – compatible with an approach conceiving science as a source of inspiration for the creators of fiction and technical imaginary worlds?

Design fiction and science fiction prototyping are examples of practices that use science fiction to stimulate creativity and invent utopian technologies that initiate innovation processes. The term design fiction was coined by science fiction author Bruce Sterling in 2005. It has since become a concept adopted by more and more economic actors. The specialists in this practice provide advice to companies, proposing various approaches, such as the invention of technological fictions to detect promising and innovative concepts. Science fiction culture is common to a large number of actors, especially among engineers, a professional category of particular interest to us here. The sociology of science and organizations, as well as the sociology of the imaginary world, has in recent years provided knowledge on the interactions between fiction, technology and R&D. This book will address a large number of theories on the contribution of science fiction to the imagination of engineers and technoscientific innovation.

In Chapter 1, Nadine Boudou explains the role of science fiction films in which technical progress has led to catastrophic situations. The post-apocalyptic genre is more and more influential in the technical imagination. The author is also interested in the *Real Humans* series (*Äkta människor* in Swedish), which shows the dangers of a massive commercialization of humanoid robots in society. The imaginary world also reveals the fears of an author's or society's future. Contemporary science fiction thus reveals the limits of a technical progress that could lead to the destruction of humanity or even the planet. Post-apocalyptic films underline the fragility of a technological civilization that conceives of itself as all-powerful. The author is also interested in the themes developed in climate fiction, which is particularly abundant since the theme of global warming warns of the dangers of uncontrolled technical progress. Nadine Boudou explains that "The end-of-the-world hypothesis on which these scenarios are based can be useful to innovators and inspire them in their research so that they can be put to work for the common good". Science fiction is indeed a powerful criticism of the impact of technosciences on ecosystems and human societies. The use of philosophy is useful, as it enables a virtuous dialogue to be established between engineers and innovators, and the rest of humanity worried about its future.

In Chapter 2, the contribution of Marianne Chouteau and Céline Nguyen questions the role of science fiction in the development of technical culture. This is presented as indispensable in order to curb the possible excesses of a society increasingly dominated by technologies that are sometimes frightening because of their potential dangerousness for humankind. The two authors share their experience as teachers in an engineering school, where they propose different approaches to students using their imagination. These are often influenced by science fiction in the form of novels, films or series. In particular, these stories help to stimulate ethical reflection on the integration of innovations into society. Thus, science fiction is at the service of technical culture. A few years ago, the two authors carried out particularly enlightening research[1] on the uses of science fiction in an engineering school, and this chapter proposes its enrichment.

Chapter 3, written by Florin Alexa-Morcov, recalls the emergence of designers, guarantors of the aesthetics of products and their relations with engineers with a specific rationality. The search for the new is seen as an important driving force of innovation. To make a product desirable, however, engineers and designers must work together. Through several examples of successful products, Florin Alexa-Morcov shows that "new objects and their representations, presented as a consequence of the innovation process, in the framework of 'performative utopias' are designed to make consumers aware, at least in part, of the hidden dimension of the innovation process". Designers are presented as magicians who end up generating a system of objects that are the source of desires guided by a fictional dimension that is often hidden or imperceptible.

In Chapter 4, Nicolas Minvielle, Remy Hemez and Olivier Wathelet, specialists in design fiction, present a study of the representations of foot soldiers in more than 300 works of fiction. They question the role of these narratives in the military sector. While the imaginary world allows us to be projected, it is also influenced by specialists in certain sectors such as the army. For some years now, design fiction has become a method of stimulating creativity in the strategic departments of institutions and even some companies. In addition, authors and directors of science fiction novels and movies use experts to provide realistic representations of future technologies. The Hollywood industry is thus at the service of storytelling that is particularly useful to engineers and designers in inventing tomorrow's innovations. The authors also propose a reflection on the methodology to be put in place to study and map imaginary worlds that have the capacity to "constitute a laboratory for projecting credible and relevant visions of the future".

Chapter 5 focuses on the Martian imaginary world. The space sector is very much influenced by science fiction, which also draws much of its inspiration from the scientific discoveries of major agencies such as NASA

1 CHOUTEAU M., FAUCHEUX M., NGUYEN C., "Les élèves ingénieurs à l'épreuve de la fiction. Quelles relations entretiennent-ils avec les mondes scientifiques et techniques fictionnels?", *Les Enjeux de l'Information et de la Communication*, vol. 16/3A, pp. 69–82, 2015, available at : http://lesenjeux.u-grenoble3.fr/2015-supplementA/22-Chouteau-Faucheux-Nguyen/index.html.

and ESA. The imaginary world of different actors from the Mars Society, founded by the American engineer Robert Zubrin, as well as the influence of science fiction on the discourse of the entrepreneur Elon Musk, will be discussed. The project for the colonization of Mars is mobilizing more and more players in a space sector undergoing rapid change, the role of the major agencies being relativized in favor of a private sector that is increasingly aware of the economic prospects of space. While the conquest of Mars was a subject long confined to communities of engineers and science fiction fans, this imaginary world project could well transform into a technopolitical ideology called *Marsism*.

Marie-Luc Arpin, Corinne Gendron, Nicolas Merveille and Jean-Pierre Revéret, in Chapter 6, question the influence of science fiction on scientific projects, but also on foresight activities. This imagination is considered a form of utopianism, through analyses inspired by Pierre Musso's research. The authors believe that science fiction must be involved in the training of engineers in order to ensure a controlled deployment of techno-utopia, of which the technical imagination is a manifestation. Thus, these works of fiction reveal their political dimension in an increasingly technical society. By envisaging in particular the emancipation of oppressed social categories, the genre ensures the development of a subversive and innovative vision of the world. Engineers are an increasingly influential social category and science fiction is emerging as a means of developing a technoscientific consciousness contributing to their identity as scientists, but also as futurists. Women's and feminist science fiction is presented as an abundant source of innovative concepts, social, political and technological at the same time.

In Chapter 7, the aim is to present an approach to innovation directly inspired by science fiction and which recently emerged in the late 2000s. The idea is to use the technical imagination to create fictions or prototypes, which R&D engineers and strategists can integrate into their research and forward-looking discourse. More and more companies and organizations are taking an interest in this innovative practice, the term of which was coined by science fiction author Bruce Sterling in 2005. The presentation then proposes a reflection on the technotypes, or technological archetypes, which contribute to the creation of the utopian technologies that are very present in science fiction novels. Reflecting on the sociological and psychological processes involved in the creation of the technical imagination is an important issue at a time when organizations are increasingly interested in design fiction that could enable them to create institutional science fiction narratives that can be used at different levels of innovation processes.

Finally, Sonia Adam-Ledunois, Claire Auplat and Sébastien Damart ask this question in Chapter 8. They start from the premise that science fiction is the result of a fascination with science, technology and innovation. They analyzed the production of dozens of critical science fiction works. References to the work of specialists in the study of science fiction provide a theoretical and bibliographical approach to the subject and help guide reflection on the functioning and influence of science fiction on society in general and on the scientific community in particular. The authors deployed a methodology called systematic literature review (SLR). In order to study academic articles dealing with science fiction, they used and analyzed results from two databases: EBSCO Business Source and JSTOR. The various arguments developed in the academic literature, and the bibliography used in this chapter will allow the reader to orient toward publications offering an often original and relevant vision on the interactions between science, science fiction and innovation.

1

Technological Innovations in the Post-Apocalyptic World: Lessons Learned from Science Fiction Movies

1.1. Introduction

Science fiction intrigues a world in which material reality and social life are disrupted by technological innovations. These transformations are characterized by advances that make it possible to relieve humans of a large number of tasks, to the point of freeing them from their constraints. In addition to being freed from all physical and intellectual effort, humanity would no longer have to endure illness and death because of therapies that would ensure health, longevity and even immortality. If we are able to escape the limits of body and mind, humanity could also escape the limits of time and space by leaving Earth. These are the contributions that some innovations promise and that many science fiction stories develop. Yet, despite the high level of technology that humanity has achieved, these scenarios continue to depict catastrophic situations as direct or indirect consequences of these innovations. The ideal level of progress based on unlimited growth and infinite innovation is contradicted by the collapse of civilization. The progress that these innovations seem to contain is limited by the apocalyptic or post-apocalyptic consequences described in these works of fiction.

While apocalyptic movies look at the before or after of a disaster, post-apocalyptic movies represent the post-disaster world. These works of fiction

---

Chapter written by Nadine BOUDOU.

all have in common a question about the future of the Earth and humanity in the event of the collapse of civilization. Depictions include Earth having lost its habitability, and humanity being deprived of the systems of protection and organization that framed and regulated its activities. These works of fiction can, on the other hand, differ on the causes of the disaster: domination of machines, pollution, pandemics, climate change and scarcity of resources. In spite of these singularities, apocalyptic and post-apocalyptic movies follow the same master plan that gives this subgenre of science fiction its homogeneity. The lessons that we can learn from these movies will be accompanied, in parallel, by those brought to us by the analyses of certain researchers who insist on the need to respect planetary limits for present and future generations.

We will first see how science fiction normalizes the idea of a world revolutionized by the use of digital technology and builds the vision of a world dominated by machines with autonomy. However, despite the description of the possibilities offered by these innovations, we will show that many movies are interested in what can result in disaster for our societies. This will allow us to analyze what these science fiction movies can be used for, since, although they allow us to imagine certain possibilities, they show their dangerousness as well as their unsustainability in the long term. Despite the catastrophic nature of these movies, we will show, nevertheless, that they can also have a positive impact on the audience.

1.2. The future machine of humanity

The Swedish television series, *Real Humans* (Lars Lundström, 2012–2014, *Äkta människor* in Swedish), raised some questions on the intrusion of robots in human societies. Over two seasons, this series developed questions about the definition of a human and what differentiates it from the machine. The story opposes humans to hubots, a contraction of human and robot, which can be humanoid robots, clones or augmented humans. Among humans, some humans form special relationships with hubots. Others reject them until they want to destroy them. The former worship technology, its innovations and the infinite possibilities it offers to the human race. The latter see it as a violation of the laws of nature and in artificial intelligence and its applications a threat to the human condition. This opposition makes it possible to show what is at stake in the debate between technoprophets and technophiles on the different conceptions of humanity, its definition and its destiny.

Dr. Eischer, the designer of hubots, belongs to the category of technoprophets and sees humanity as an infinite field of exploration and experimentation. The accidental death of his wife and son accelerates his research and applications through the creation of a being half-human, half-robot, a cyborg. His son, Leo, becomes the most accomplished representative of a cyborg. As for his wife, Bea, her spirit is transferred into a hubot body, an android robot. Unlike other hubots, it is a free robot that has benefited from programming based on a source code that integrates the transgression of Asimov's laws. Dr. Eischer has developed, through a few hubots he calls "his children", the code that ensures their autonomy and awareness of their condition. These hubots, defining themselves as free, do not hesitate to commit murder to recharge their electricity and express their intention to replace humans they consider inferior to them. They are able to reset themselves, and download all programs to increase their brain power and physical performance. They repair themselves and can transfer their mechanical brain into a new hubot body.

The benefit for everyone lies in the autonomy given to all hubots and the possibility for humans to increase their capabilities. The emotional reactions of humans and their slow learning and execution are replaced by a new form of life based on programming, information and digitization and their brains reduced to a program devoid of reflection, intuition and imagination. Hubots are traded and trafficked from the virtual machine to the sex object. They perform all the tasks for which they are programmed, without getting tired, without questioning and without any hesitation. They give hope to the humans around them to achieve such performances by transforming themselves into hubots. The issue of the series is then, for humans and for the few liberated hubots, to find the code that Dr. Eischer, before his death, made sure to hide because he considered it too dangerous to leave in the hands of humans or hubots. Having conceived this type of creature, the scientist, like Dr. Frankenstein, realizes the dangerousness and monstrosity of such an innovation. In this case, the trouble does not come only from a humanity dominated by its machines, which is a recurring theme in science fiction works, but from a humanity whose dream is to become a machine itself. The dystopian character of this series lies in the description of a world in which humans would no longer be evaluated solely on the basis of speed and efficiency.

Cédric Biagini describes the fascination of new technologies as follows: "With them, everyone thinks of freeing themselves from the constraints and limits of time, space, human relationships and their bodies... They plunge the individual into excess" [BIA 13, p. 362, author's translation]. These possibilities would render obsolete the laws of nature, which would be replaced by a new way of being, designing and acting. Yet, far from remedying the presumed inadequacies of our condition, technological prowess is proving to have destructive consequences for the environment and by extension for humanity. As Philippe Bihouix notes, "Computer science is obviously the field where the level of our ambitions is the most hallucinating, where the – usual – hypothesis of infinite resources on the planet is the most obvious" [BIH 14, p. 226, author's translation]. Despite the infinite possibilities that innovations promise us, science fiction, but also numerous analyses, show us their limits.

1.3. A pending world?

In *Real Humans*, the possession of a hubot for domestic or professional use is a sign of social distinction. This unequal situation is discussed in *Blade Runner* (Ridley Scott, 1982), *Gattaca* (Andrew Niccol, 1997), *AI* (Steven Spielberg, 2000) and *I, Robot* (Alex Proyas, 2004). In these movies, innovations in artificial intelligence, robotics or biotechnology benefit only a minority of individuals at the expense of a majority who are excluded. While privileged populations enjoy a certain comfort, others live in unsanitary neighborhoods and cities, victims of scarcity, persecution and discrimination. The authors of *La face cachée du numérique* legitimately ask: "If the 'information society' is so energy and material efficient, and generates growth for all, how is it that inequalities and consumption of resources continue to increase?" [FLI 13, p. 45, author's translation]. If everything seems to be designed, on the surface, to improve the fate of humans, in reality, this economic and technological development benefits first and foremost those who exploit it and contributes to the unbearable future of the Earth. Science fiction, in the age of the digital revolution, has noted an internal contradiction in this technoindustrial logic.

The appearance of freedom offered by these technologies becomes a mechanism of surveillance and control in both professional and private life. A movie such as *Minority Report* (Steven Spielberg, 2002) develops all the implications. Above-ground life dependent on artificial intelligence, as described in *Real Humans*, serves as a metaphor for the derealization of the world that the digital revolution is accomplishing. Éric Sadin takes the example of the television series *Mr. Robot* (Sam Esmail, 2015–2017) and writes, "Probably the contemporary individual has recently been suffering from '*Mr. Robot* syndrome', exhilarated by the impression that reality, as a field everywhere strewn with constraints, is eclipsed when it enjoys 'computational facilitation'" [SAD 16, p. 204, author's translation]. This state of dependence on computer imperialism is described by François Cusset as follows: "Such are the ravages of abundance: loss of the effective experience of life, companion items, precluded desires, access to everything at once, and their effects on social relations and on the future of the world" [CUS 18, p. 162, author's translation].

Large-scale manufacturing of hubots can only generate an additional accumulation of non-recyclable waste. Movies such as *Soylent Green* (Richard Fleischer, 1974), *AI*, *WALL-E* (Andrew Stanton, 2009) and *Elysium* (Neill Blomkamp, 2013) have addressed this subject. The technoindustrial system generates pollution as much for the extraction of raw materials, their transformation as for their production and then their use and storage in the form of waste. Marc Atallah writes, "*WALL-E* reminds us that our comfort-obsessed world is a self-sustaining system, since the disposal of the waste we constantly produce also involves waste-producing logistics" [ATA 15, p. 924, author's translation]. These works of fiction demonstrate that innovations due to the digital revolution are not necessarily emancipatory, ecological or sustainable in the long term. According to Guillaume Pitron, "The so-called happy march into the age of dematerialization is therefore a vast deception, since it actually generates an ever-increasing physical impact" [PIT 18, p. 68, author's translation]. The synergy of the ecological, climatic and geopolitical crisis described in post-apocalyptic narratives highlights the incompatibility between the perpetuation of our lifestyles and the reality of planetary limits. These fictions give a representation of the great shift that many analyses predict and make humanity a species threatened by the endangerment of the ecosystem on which it depends.

The dependence of hubots on electricity reveals their vulnerability and suggests, as in *Mad Max* (George Miller, 1979-2017) or *Matrix* (A. and L. Wachowski, 1999-2003), the violence of a world in which energy sources are becoming scarce. Moreover, the depletion of the rare metal resources needed to manufacture these machines is pushing earthlings, as in *Avatar* (James Cameron, 2009), to move to colonize new territories by expropriating or exterminating their inhabitants. The ecological and geopolitical violence generated by the exploitation and depletion of resources is indeed shaping our common future. The fact that in *Real Humans* the group of free hubots, after finding the source code and after releasing the other hubots, are planning to suppress the human species heralds a post-apocalyptic future for the human species. This demonstrates the irrationality of innovations whose purpose would be to make us disappear. James Cameron in *Terminator* (1984) had already addressed this theme by imagining the threat that robots created by a supercomputer could pose. Apocalyptic and post-apocalyptic works highlight the absurdity of a development model that, while it is not viable in the long term, is responsible for a planetary catastrophe.

1.4. Consuming the world

From *Soylent Green* to *The Road* (John Hilcoat, 2009), *Wall-E, Avatar* or *Interstellar* (Christopher Nolan, 2014), science fiction movies evoke, through a dying Earth, the violence generated by our world. These movies express a contrast between the high level of technology achieved by the human species and the state of the planet. While, in these fictional works, everything seems possible from a technical point of view – cloning or robotizing human beings, humanizing machines or colonizing space – they show populations whose living conditions do not appear to have improved as a result. On the contrary, the world they live in is unequal, violent, polluted and inhospitable. The benefits that humanity derives from the technical administration of life and natural resources cannot indefinitely mask their disadvantages and their incompatibility with the perpetuation of life on Earth in good conditions. Other movies echo climate change like in *The Day After Tomorrow* (Roland Emmerich, 2004), ecological disasters like in *2012* (Roland Emmerich, 2009), wars leading to technological and ecological disasters as in *Mad Max*, *The Book of Eli* (Allen and Albert Hughes, 2010), transformations following manipulations on living things as in *The Planet of the Apes* (Franklin J. Schaffer, 1968), *I Am Legend* (Francis Lawrence, 2007) and *28 Days Later* (Danny Boyle, 2003). These movies are a metaphorical translation of our anthropological and sociopolitical reality, which François Cusset describes as follows, "for while industry is ravaging entire regions, while agribusiness is poisoning the countryside, and while Promethean modernity is destroying the biosphere from within, the big cities are becoming less and less habitable" [CUS 18, p. 71, author's translation]. The belief in the exorbitant powers attributed to techniques comes up against criticism based on studies and prospective studies that show their systemic limits in a finite world. Guillaume Pitron notes that "we are going to consume more minerals in the next generation than in the last 70,000 years, that is, in the 500 generations that preceded us" [PIT 18, p. 214, author's translation]. And, "to manage the data that flows through and to operate the cooling systems, a single data center consumes as much energy every day as a city of 30,000 people" [PIT 18, p. 67, author's translation]. According to a report on global inequality, coordinated in particular by Thomas Piketty, published on December 14, 2017, "since the 1980s, the richest 1% have benefited twice as much from income growth as the poorest 50%. And for those in between, incomes have either stagnated or declined"[1].

These works of fiction show that technoindustrial progress, if it is exclusively at the service of power, profit and domination, cannot guarantee real progress. Despite the allure of a world subject to new technical standards dominated by generalized robotization, the continuous connection and apparent liberation of humanity, its unsustainable character emerges. The apocalyptic and post-apocalyptic imaginings coincide with the idea that if our development model depends on non-renewable resources and the exploitation of rare metals, if it generates significant energy expenses and causes the release of polluting and non-recyclable substances, it condemns those it is supposed to serve to widespread scarcity. What in all the fictions devoted to these issues leads to a liberticidal and deadly world in which injustice and insecurity reign, from which barbarism arises. Since 1968, George A. Romero's movies have popularized zombie invasions as a metaphor for the consumption of the world and the loss of what Renaud Garcia calls "living subjectivity" [GAR 18, p. 205, author's translation]. This is why "the imaginary world of technical evolution should unfold in a universe other than the one already there and largely drawn by technical macro-systems" [GRA 13, p. 44, author's translation]. This imaginary world of technical innovation is still to be constructed.

While these movies in no way call into question the real progress made by technoscientific innovations, they do insist on the potential risks they contain. While the science fiction author Neal Stephenson has reservations about the relevance of catastrophic stories, criticism of innovations and their consequences should be free to be exercised. What a writer such as Paolo

1 Available at: https://www.lemonde.fr/les-decodeurs/article/2017/12/14/les-inegalites-dans-le-monde-en-hausse-depuis-quarante-ans_5229478_4355770.html.

Bacigalupi, a representative of climate fiction, claims. While, as he writes in an interview, "scientists, thanks to their research, allow us to anticipate certain scenarios", the role of fiction is "to go further than them, by painting a visceral image of what is likely to happen, based on these data"[2]. These works of science fiction conceived as safeguards serve as a counterbalance and reveal their usefulness in maintaining public vigilance.

1.5. A finite world

Science fiction recognizes the power of technology, but in many of its scenarios it includes descriptions of degradation committed in the name of development that does not take into account the real needs of humanity or the long-term preservation of the environment to which it belongs. Movies such as *Blade Runner, Gattaca, AI* or *Minority Report* show how the dangerousness of the digital hold results in a large-scale robotization of social organization, generalized control and impoverishment of lifestyles and subjectivity. Miguel Benasayag notes that, "we are thus surrounded by contemporaries who, in an almost pathological misunderstanding of their lives and the world, are qualified as 'intelligent' because they are capable of functioning like machines (...) by producing deterritorialized and decontextualized algorithms" [BEN 15, p. 77, author's translation]. The problematic nature of our situation should rather encourage us to seek solutions by staying connected to the immediate reality. The interest of post-apocalyptic movies is to warn of the fragility of a technoindustrial system dependent on natural goods that cannot be mechanically reproduced. Through them, as in *The Road*, we can envisage the disappearance of a world that no longer has sufficient conditions to ensure its reproduction. Without being predictions, these works are, as Marc Atallah writes, "scenarios, that is, in short, models, aimed at making our lives intelligible over time" [ATA 15, p. 924, author's translation].

As Ulrich Beck analyzes in *La société du risque*, our societies are placed in a new situation because of the risks associated with modernization, which are amplified by its development. These threats, which were specific to the 20th Century and which are becoming clearer in the 21st Century, feed an imagination that is echoed in many movies. They generate concerns that

2 Available at: https://usbeketrica.com/article/cli-fi-des-fictions-pour-prendre-conscience-du-peril-climatique.

spread beyond the country that produces these movies and that develop upstream of movie production. It is a reversal that humanity should be facing today. The meeting of the Greek elements *kata*, "down" and *strophê,* "the reversal" gave rise to the word catastrophe. Our concern for nature prompts us to turn our attention to it as a source of future disasters. The depletion of energy sources, metals and minerals, the destruction of biodiversity or climate change are concerns that are widely expressed in public life. In a column published in the French newspaper *Le Monde* on June 26, 2017, Stéphane Foucart wonders what part the ecological crisis has played in the war in Syria or in the exodus of migrants trying to cross the Mediterranean. He writes, "Without succeeding in reconnecting recent history and current events to changes in the environment, we are condemned to remain blind to what promises to be one of the great decisive factors of our collective destiny"[3]. These concerns form the framework of many movies through which a world view of the precarious state of the planet is generalized. They give rise to a sense of urgency that is reflected in the acceleration of the process that movies such as *The Day After Tomorrow*, *2012*, *The Road*, *Mad Max: Fury Road* or *Interstellar* bring to the stage. They are not creators of a world view but prolong a mental climate that is constructed independently of their influence.

Stanley Cavell considers that cinema can "contribute to the education and intelligence of a culture, or let us say to the understanding that a culture has of itself" [CAV 03, p. 13, author's translation]. In these movies, the fears of the beginning of the century crystallize. They lie in the difficulty of projecting oneself into a future that guarantees its continuity. Anthony Giddens, who analyzes the institutional transformations linked to modernity, notes: "The lack of control which many of us feel about some of the circumstances of our lives is real. It is against this backdrop that we should understand patterns of privatism and engagement. A sense of 'survival,' in Lasch's use of this term, cannot be sent from our thoughts [...]" [GID 90, p. 152]. Hence all the talk about sustainability. If nothing seems to be sustainable anymore, how can we look to the future? Such is the main questioning that guides these stories dominated by the impermanence of resources and social systems that, in the 1950s or 1960s, cinema could not yet thematize.

---

3 Available at: https://www.lemonde.fr/climat/.../la-pedagogie-des-canicules_5150991_1652612.html.

Fictions are an act of unveiling and pointing out what in our present situation increases danger. Their objective is not to question the progress that these tools bring, but to insist on the constraints of reality and the limits of our environment. Eco-fictions or post-apocalyptic movies constantly dramatize and describe the consequences of the misuse of nature due to massive and irrational exploitation of its resources. By imagining the worst that our world seems to contain, these fictions accompany, narrate and give meaning to the many concerns that the current situation maintains. Marc Atallah writes, "At a time of global warming and various climate issues, it therefore seems imperative to appropriate scientific theories that, although relevant, are nonetheless devoid of anthropological significance" [ATA 15, p. 924, author's translation]. The post-apocalyptic genre, by representing the risks contained in our current mode of development, acts as a "fire alarm". As Walter Benjamin wrote in *One Way Street*, "The burning fuse must be cut before the spark reaches the dynamite" [BEN 28, p. 157].

Peter Sloterdijk thus formalizes the maxim of action that must be the maxim of our times: "Act in such a way that the practice of plundering and outsourcing that has been in force until now can be replaced by an ethos of global protection" [SLO 11, p. 63, author's translation]. The power structures that guaranteed the old order are cracking and science fiction is becoming the instrument of a paradigm shift. This theme of a planet that now presents itself to us as finite and limited echoes a remark by Michel Serres in *Temps des crises*. He writes: "We thought, bravely, that all our history consisted in constantly fighting against a force ever higher and deeper than our own. The image is reversed: we now know that we are infinite, of reason, of research, of desire and will, of history and power, even of consumption, and that nature, facing us, is finite" [SER 12, p. 48, author's translation]. This vision of the world is not limited to a country's borders and is becoming more widespread. This cinema of the beginning of the 21st Century is at the crossroads of a new world, which is becoming aware that the future is still to be invented because we can no longer project in it the repetition of what was before. By acknowledging our defeats and our powerlessness in the face of a runaway mechanism, many movies sign the bottom of a page in the history of humanity that is turning. It must renew itself and its relationship with its surroundings, so as not to disappear.

1.6. Conclusion

Jean-Marie Schaeffer considers that the fictional universe "participates in my learning of the world, in a kind of dialectical oscillation between its capacity for representation and its repulsive function" [SCH 99, p. 38]. The pleasure of watching stories, sometimes charged with violence, comes from the fact that they manage to hold our attention and interest. They have the power to refer us back to what we judge to be the state of our world and the effect that state of the world has on us. By imagining the worst of what could happen to our world and our humanity, post-apocalyptic fiction can serve as a warning. As Michael Fœssel puts it, "The most regularly cited positive consequence of catastrophism is that it contributes to the emergence of a global consciousness" [FŒ 12, p. 243, author's translation]. Through the reality that is reflected in it, fiction can enunciate and denounce something at work in today's world. As Jean-Michel Valantin notes, "This threat posed by the shock of the Anthropocene is dealt with in an almost industrial manner by American and British cinema and series" [VAL 17, p. 284]. Through a playful pretense, the audience immerses itself in these stories and its involvement nourishes its understanding and interpretation of the real world.

By alerting or alarming us about a possible future of the Earth and humanity, post-apocalyptic fiction lays bare our doubts and concerns about the state of our world. The catastrophism of these movies cannot be reduced to a simple fascination with disaster. On the contrary, it can accompany "the democratic debate by helping it to break with the *continuist* myths that now structure the dominant political imagination: irreversible development, eternal growth, perpetual progress, etc." [SEM 15, p. 143, author's translation]. What is thus revealed through these works and the aesthetics of disaster that characterizes them is a meeting point between imaginary worlds of disaster and a sense of global risk. This feeling is reinforced by the analyses of many researchers, relayed by the media, according to which a change in the development model is imperative for us. The staging of the Apocalypse can be a reference to a desire for change. The end-of-the-world hypothesis on which these scenarios are based can be useful to innovators and inspire them in their research so that they can be put to work for the common good. Can the horrific and hallucinatory spectacle of the collapse of the civilized world and its consequences not, if it can be received from afar and jubilation by the public, give rise to a strong hope for change, so that the worst described in these fictions cannot become reality?

1.7. References

[ATA 15] ATALLAH M., "Science fiction", *Dictionnaire de la pensée écologique*, PUF, Paris, pp. 922–925, 2015.

[BEC 01] BECK U., *La société du risque. Sur la voie d'une autre modernité*, Flammarion, Paris, 2001 (1986).

[BEN 15] BENAYASAG M., *Cerveau augmenté, homme diminué*, La Découverte, Paris, 2015.

[BEN 28] BENJAMIN W., *One-Way Street,* Belknap Press, Paris, 1928.

[BIA 13] BIAGINI C., *L'emprise numérique*, Éditions l'Échappée, Montreuil, 2013.

[BIH 14] BIHOUIX P., *L'âge des Low Tech*, Le Seuil, Paris, 2014.

[CAV 03] CAVELL S., *Le cinéma nous rend-il meilleurs?*, Bayard, Paris, 2003.

[CUS 18] CUSSET F., *Le déchaînement du monde. Logique nouvelle de la violence*, La Découverte, Paris, 2018.

[FLI 13] FLIPO F., DOBRE M., MICHOT M., *La face cachée du numérique*, Éditions l'Échappée, Paris, 2013.

[FŒ 12] FŒSSEL M., *Après la fin du monde. Critiques de la raison apocalyptique*, Le Seuil, Paris, 2012.

[GAR 18] GARCIA R., *Le sens des limites*, L'Échappée, Paris, 2018.

[GID 90] GIDDENS A., *The Consequences of Modernity*, Stanford University Press, Palo Alto, 1990.

[GRA 13] GRAS A., *Les imaginaires de l'innovation technique*, Éditions Manucius, Paris, 2013.

[PIT 18] PITRON G., *La guerre des métaux rares. La face cachée de la transition énergétique et numérique*, Les Liens qui Libèrent, Paris, 2018.

[SAD 16] SADIN É., *La siliconisation du monde. L'irrésistible expansion du libéralisme numérique*, Éditions L'Échappée, Paris, 2016.

[SCH 99] SCHAEFFER J.-M., *Pourquoi la fiction?* Le Seuil, Paris, 1999.

[SEM 15] SEMAL L., "Catastrophisme et démocratie", *Dictionnaire de la pensée écologique,* PUF, Paris, pp. 141–144, 2015.

[SER 12] SERRES M., *Temps des crises*, Le Pommier, Paris, 2012.

[SLO 11] SLOTERDIJK P., *Le monde n'a plus de temps à perdre. Appel pour une gouvernance mondiale solidaire et responsable,* Les Liens qui libèrent, Paris, 2011.

[VAL 17] VALANTIN J.-M., *Géopolitique d'une planète déréglée*, Le Seuil, Paris, 2017.

2

Using Science Fiction in Engineering Education: Technological Imagination as an Element of Technical Culture

2.1. Introduction

The general context of the beginning of the 21st Century means we are interested in and question technology more than ever. Scarcity of resources, global warming, accentuated artificialization of the world, digital development, demographic growth, destruction of biodiversity, etc., are all subjects in which human beings and the technology[1] they produce are involved. These major contemporary issues allow us not to separate technology from society: "What is at stake is fundamental not only because the techniques increase the capacity for action but above all because they give the capacity to modify the capacity for action of man" [VIN 88, authors' translation].

At the center of this context are the engineers. As designers or makers, they shape the world with their technological achievements. It therefore seems natural to us that they should be able to take a critical look at both their jobs and their productions. This is not a matter of making a speech against engineers. But rather highlighting the need for them to understand what they produce, here on a symbolic level, through the part of humanity

Chapter written by Marianne CHOUTEAU and Céline NGUYEN.

1 Technology understood in the sense of the processes and the technical objects resulting from these processes.

that is involved in the technology. In other words, it is a question of bringing technology back into the circle of culture by revealing its cultural part. Technology as a human activity of object production is indeed inserted in a web of symbols, imaginary worlds, beliefs and cultural references. To do so, we will make a detour through the technical culture of fertile ground, which is essential to the understanding of technology and its production. A number of authors from different disciplines such as André Leroi-Gourhan, Jocelyn De Noblet, Philippe Roqueplo, Jacques Perriault and Bruno Jacomy have campaigned for the development of a "technical culture" in France. Their main argument was to say: "We therefore believe that a technical culture is necessary because it can be defined as the possession of a minimum amount of knowledge and know-how allowing the reappropriation of our environment. We mean that it is one of the conditions from which it is possible to appropriate technology" [DE 81, authors' translation].

Our approach is an extension of these militant initiatives developed in the 1980s in France, but in a particular context, that of training in the engineering profession, a key moment in an engineer's life[2].

Indeed, in this contribution, our aim will be to show that science fiction is a central element of technical culture in that it is a particularly enlightening cultural genre appreciated by engineers. Science fiction makes it possible to understand, through representations, imaginary worlds, ideologies, etc., that it conveys what socially and symbolically inscribes technology as an activity and the technical object as a result. We want to show that works of science fiction are an interesting lever to nourish, work and make visible the engineer's technical culture and that, in doing so, it has its full place in an engineering school education, that it is not simply a form of "open-mindedness" or "the icing on the cake".

2 We are aware that there are many different kinds of engineers and that the profiles and specialities (R&D, sales, design, etc.) of each are sometimes very different. The definition proposed by the *commission des titres de l'ingénieur* (French commission responsible for the evaluation and accreditation of higher education systems training engineers) seems to us to be appropriate: "the engineer's job consists of posing, studying and solving, in an efficient and innovative way, often complex problems of creation, design, realization, implementation, control of products and systems or services, possibly their financing and marketing within a competitive organization. It integrates concerns for the protection of humanity, life and the environment and, more generally, for collective well-being" (References and orientations of the CTI, 2015, authors' translation).

We are therefore well aware that this chapter is about engineering students and engineering education and not about engineers' imaginaries. We do not measure the impact that this kind of training can have on their professional practice and how they reinvest their imaginary when they are in business, when they leave school or later on. In this sense, this chapter is not the product of a survey or an assessment whose results would be given, but rather a proposal in action that attempts to respond to the challenges they face today: to conceive ethically, to prioritize societal issues, to be creative, etc., which in our opinion requires a technical culture.

So, in this contribution, after defining technical culture as we understand it, we will draw up a brief overview of science fiction in order to quickly retrace its history and uses. We will show how it is part of the technical culture. Finally, through examples of pedagogical modules developed in engineering schools, we will highlight how technical culture comes to engineering students through science fiction.

2.2. What is technical culture?

2.2.1. *In the name of autonomy*

Generally, technical culture is about understanding how a technical object works but also about grasping its symbolic significance. André Leroi-Gourhan, in the foreword to the *Manifeste pour le développement de la culture technique*, testifies that the "function-form" relationship, otherwise known as "design", is "one of the privileged paths of technical culture" [LER 81, authors' translation], referring to a culture close to design. The anthropologist incorporates, in a second movement, "aesthetics" and "magical-religious institutions", which allows us to understand the technical object's cultural value through the action it makes possible. Thus, culture enters into the technical object, which in turn produces culture.

As early as the 1950s, the philosopher of science and technology Gilbert Simondon had already noted that:

> Culture is unbalanced because it recognizes certain objects, like the aesthetic object, granting them citizenship in the world of significations, while it banishes other objects (in particular

> technical objects) into a structureless world of things has no signification, but only a use, a utility function. [SIM 12]

Later, technical culture was the subject of particular attention in the 1980s, particularly in an attempt to counter the ostracism to which it was subjected, at least in France, in the field of legitimate culture. Jocelyn de Noblet, for example, draws our attention in a manifesto for technical culture, at a time when "endless progress" was being questioned and its harmful effects were being attested [DE 81]. The aim was to take the measure of the technical systems in which technical objects are embedded, and to go beyond mere familiarity with technical objects to acquire knowledge about technical objects. It was aimed more at citizens: "Technical culture will be for us the thinking that is at work when a technology does not escape its users" [DE 81, authors' translation]. This technical culture aims to escape a double alienation: that of not mastering one's environment and of depending on organizations that have the skills. For this author, therefore, it aims to encourage autonomy.

Philippe Roqueplo's point is not far off:

> The technical culture consists of the possession of knowledge and know-how likely to base a minimum of personal control over our environment and control over the activity of those whose competence proves to be indispensable. [ROQ 83, authors' translation]

To this end, three definitions of technical culture can be distinguished according to the point of view adopted: how it was made; how it is used; when and why it was invented [JAC 93]. Each of them aims to appropriate the technical object in question.

Yves Deforge, for his part, allows us to think that the technical culture

> requires an objective and operational knowledge of the technical (and scientific) world but also, beyond that, an aptitude for symbolic and universal, that is to say, properly philosophical thinking. [DEF 96, authors' translation]

Thus, the author hopes that approaches that are often opposed when it comes to technology will interact and that the symbolic part will nourish the action to form a cultural whole:

> What we are suggesting is to put technology and science under the control of a culture that is farsighted and active. [DEF 96, authors' translation]

In 1993, Bruno Jacomy defined more precisely what the engineer's technical culture could be, making it a set of knowledge related to technology and the technical object: knowledge of the history of technologies and industry, the history of the engineer and the innovation process. He favored a cultural approach to technology: every object being representative of a history, a system and part of a particular socioeconomic and political environment with the aim of innovating "better" and developing "appropriate technologies" [JAC 93].

2.2.2. *For a non-segmented technical culture*

Starting from this fertile ground, we proposed to define a technical culture [CHO 17], which is addressed to a diversity of actors including engineers. This definition is based on three complementary levels of analysis.

The first level concerns calculations, materials and properties of the object. It is scientific and technical knowledge that materially defines the object. It is the level of knowledge related to the structure, the composition of the object and the functions; it is the level of technical design and the level of engineering sciences classically defined. This level goes without saying in an engineering school.

The second level is the use of the object. It shows how our relationship with the object has a "social density" [JOU 00] when we use it. The use depends on our projects, our lifestyle, our age, our norms and values, our activity, etc. It modifies, accentuates practices and allows us to reveal the logic of use [PER 98], strategies and tactics [DE 90, JAU 11]. This level thus makes it possible to account for the user's activity, for everything that is at stake in their relationship with the object. For example, it is here that we can report on the very rapid adoption of motive by divorced parents and their children, allowing them to develop conversations in complete autonomy and freedom.

The third level concerns the more global meaning of the object. It is symbolic, political, ideological, etc., knowledge that enables us to understand the object and the changes it brings about on a wider scale. It places the object within the major contemporary issues: waste reduction, access to health care, education, etc. This last level is political in nature because it allows us to think about the direction that technology gives to the society it helps to shape. It takes place on a civic scale, allowing for debate on technologies.

The technical culture as we understand it implies that these three levels are not segmented and especially that they are not worked on separately. A designer (engineer or not) can also reflect on the meaning of the object from the citizen's or user's point of view. A user can become a designer and question the political meaning of their practice. This view of things is not necessarily taken for granted in an engineering school. Indeed, while the first level is easily accessible for student engineers, the second and third levels are not so easily accessible. Our aim is therefore to ensure that engineering students (and engineers afterwards) are aware of these three levels so that they set up a global reflection articulating them. In other words, they are able to think together about the functions and materiality of the object, its uses and its more political and ideological implications.

2.3. Science fiction, technology and narrative: fertile connections

Science fiction conveys representations, promises or apprehensions about technology, and in this way, it participates in the symbolization of technology. It then constitutes one of the ingredients of technical culture as we have just defined it by being situated at three levels. It has other assets: science fiction is popular among engineering students. It is part of their culture. Although not all of them are great readers – there are, however, a few regular readers among them – the vast majority of them are spectators (film or TV series) [CHO 15a]. Moreover, science fiction – whatever the medium through which it is disseminated – offers the possibility of making a detour, of not entering directly into the questions but of approaching them by other means. It allows engineering students to invest in the technical culture.

2.3.1. *Science fiction, a sociotechnical genre*

We will not give a complete and exhaustive history of the genre here, but it should be recalled that one of its functions, at its beginnings in the 1920s, was, according to its inventor Hugo Gernsback, that of a tool for popularization, a genre designed to encourage vocations among scientists and engineers [CON 12]. He gambled that by showing possible technological worlds, science fiction could provoke desire. Many authors have looked at the contributions of science fiction. One of the most often mentioned is that of expressing desires and fears: desire for a better society, and fear of a dystopian world. The fictional nature of the stories and characters is even an asset:

> Romantic characters have a better grasp of the truth of a period than real characters do as understood by sociologists or historians. [TOF 17, authors' translation]

Pierre Musso defines science fiction as "a literary genre contemporary with the industrial revolution, which seeks to conceive a different world, even to predict the technical and social future. Born in a period of faith in progress, science fiction is now marked by questions about the harmful consequences of certain innovations" [MUS 05, authors' translation]. Echoing Gilbert Hottois [HOT 00], he argues that contemporary science fiction has indeed lost its modern "idealism" to "exploring human society" and becoming part of postmodernism.

In the first part of *Fabriquer le futur*, Pierre Musso proposes to use the imaginary to show that innovation is also a "historical and social construction" [MUS 05] and not just a production that goes without saying, without a cultural basis. The author shows very well how the imaginary world is central in the design process (central stage of the innovation process) via the plurality of imaginary carried by the actors of the process: designers but also users, large organizations, writers, advertisers and trend-setters. He shows how these mediations carry utopias, ideologies or are based on great myths, and as such, he gives science fiction a special status as a vector of social utopias.

Science fiction is also a cultural marker. Indeed, depending on the country, it does not play the same role. While science is perceived as a fascinating danger in France, the perspective is more critical in Great Britain when it comes to thinking about its use. In the United States, literature is more optimistic. The reason for this is that the authors come more from literary spheres, whereas elsewhere they navigate more in scientific ones [KLE 16], and therefore more marked by self-criticism. Moreover, imaginaries evolve over time. On the subject of the technoimaginary world, Georges Balandier [CHA 94] observes a transformation: "techno-messianic utopias" are gradually being replaced by "techno-catastrophic" utopias, a sign of a much more critical time and of this loss of idealism.

The literary or cinematographic genre of "science fiction" is special in that it makes science and technology one of its main ingredients. A kind of exploration, of impulse, of a frontier [CON 12], it is one of the most relevant vectors allowing us to explore, through fiction, our link with technology. Indeed, the imaginary worlds present in science fiction stories are largely carried by the technical objects that populate the proposed universes: space shuttles, underground habitats, flying cars, light sabers, humanoid robots, modified or augmented humans. Not only are these the markers of the genre, but they also advance the story with props or even main or secondary characters.

In doing so, the technology present in science fiction "is always a mixture of *tekhné* and *logos*" [MUS 13, authors' translation]; that is to say, it is the result of a technical act, a fabrication, and it generates knowledge and a discourse, a word. This word is that of the technical object itself, which is never silenced, but also belongs to the author and the public who receives it.

2.3.2. *Science fiction: a special genre in the service of technical culture*

In what way then does this kind of thing constitute a lever for the development of a technical culture? In an attempt to answer this question, we have identified three complementary functions that science fiction stories can play within the technical culture framework to be developed at the three levels proposed.

The first function is to encourage and surround the technology with a significant imaginary world. Science fiction, rich in technological imagination, can indeed function as a reservoir of ideas, suggestions and desires. Science fiction authors draw on current events and scientific "reality" to construct their stories. In contrast, the genre allows designers to set in motion an innovation process, either by issuing a challenge that a new technology could meet (a new way of getting around, of feeding oneself, etc.) or by more or less directly suggesting a technical solution to an identified problem. While it goes without saying that science fiction never provides a turnkey solution, it can nonetheless be used to formulate wishes, be a source of inspiration [RUM 16-17] or to launch new avenues of work.

The European Space Agency, for example, produced a report in 2002 entitled *Innovative Technologies From Science Fiction*[3], the result of a collaboration between researchers, engineers and authors who drew up a list of ideas to be explored, searched through works of fiction for inventions and innovations that could be developed and identified as viable ideas. Space flight suits were analyzed in particular and more particularly the vinyl suit or water recovery clothing in *Dune*. Much earlier, and even while alchemy was not science fiction, the latter was a driving force of knowledge about metal during the Middle Ages [GAR 05]. Closer to us and for Gerard Klein, transhumanism is clearly inspired by science fiction. One of his most fervent defenders, an engineer at Google and founder of the University of Singularity, Ray Kurzweil announced, for example, that humanity will merge with artificial intelligence in 2045 or that we will disappear, overtaken by machines[4].

The computer field – and this is not foreign to transhumanism and NBIC convergence on which it is based – seems to be marked by the imprint of a fiction that goes back a long way in time, that of "artificial creatures":

> Scientists and engineers in this field have been dreaming for more than 50 years of building an 'intelligent robot' that would be 'in the image of man'. Here the future joins past myths, and science is closely linked to the imaginary world. [BRE 13, authors' translation]

For his part, the sociologist Patrice Flichy [FLI 89] has clearly shown that the desire for universality has been a founding principle in telecommunications. Offering means of communication to anyone and

3 See, for example, https://www.thespaceoption.com/publications/IAF-01-IAA.8.2.02%20ITSF.pdf.

4 See, for example, https://www.01net.com/actualites/ray-kurzweil-espere-la-fin-de-l-humanite-pour-dans-douze-ans-1125964.html.

anywhere (i.e. the revolution proposed by cell phones) allows us, for example, to make a link with the myth of Babel. Similarly, the inventors of recording devices were driven by the desire to reproduce images and sound in series [FLI 89], a quest for immortality in a way. Finally, the author also showed very well how the history of the Internet is made of utopias and ideologies present in the minds of the founders as well as in the discourses of the media representatives [FLI 01].

In this perspective, Pierre Musso's work follows an interesting stance because it seems necessary to him to keep in mind that the imaginary is an ingredient of a successful innovation, in the sense that the thoughts of the designers needing to coincide with that of the users. The author then speaks of the "co-elaboration of representations" or the "convergence and sharing of references" [MUS 05]. His aim is to go beyond a Manichean vision of things. On the one hand, the rejection of the imaginary world by some, because it is considered irrational and therefore non-existent is observed. On the other hand we can notice an adherence or even a fascination particularly visible in "imagery produced by standardized media campaigns" [MUS 05], imagery used to facilitate the adoption of a product or service or to initiate a research project. Thus, he highlights the cultural side of any innovation: engineers and technicians are immersed in fictional, more or less standardized technological universes that feed their desires. These universes are sometimes identical to those of the users, or are opposed to each other, complementing each other, which shows that complex mechanisms are at play in the object and in our relationship with it. Thomas Michaud [MIC 14] has clearly shown the importance of the prophetic unconscious, which is very present among certain computer scientists in the Orange R&D department, such as in its managerial operation or in the marketing register.

The second function is of a more reflexive nature: it is one that allows the analysis of the representations and, *in fine*, the understanding of our relationship with technology. Works of fiction are particularly relevant for staging technical objects, uses and applications that are based on political intentions, values and social choices. In doing so, it allows us, whether we are engineers, politicians, users and/or citizens, to identify the promises and fears linked to the future of society and the technology that is part of it:

> Spectators are also citizens and they need science fiction to set a collectively shared vision of the future. [GUE 14, authors' translation]

The strength of this is therefore to encourage us to project ourselves and mobilize our imaginary to think about the outlines of the desired or feared society so as to take relevant paths and avoid pitfalls. In doing this, science fiction writers necessarily draw inspiration from contemporary society to exaggerate certain features of it. They thus provide food for thought based on socioeconomic explorations and real or underlying dysfunctions.

Here, it is not prediction that is interesting or sought after, but the understanding of society's trajectories to make proposals. In this sense, foresight remains the discipline that exploits the genre the most because "the futurist is legitimately tempted to look for leads in this continent of imaginary representations of the future that science fiction literature constitutes" [KLE 16, authors' translation].

Moreover, the technological imaginary central to the genre highlights the myths and flaws of humanity. While, as we have said, myths can inspire technical creation, they can also highlight our shortcomings, like the myth of Prometheus, which, on the one hand, helps us to understand how technology empowers man but, on the other hand, allows us to become aware of technological immoderation. And it is the interpretation of the myth of Prometheus that also changes with the times and our relationship with technology as Dominique Lecourt points out [LEC 96]. The imaginary of ICTs is largely based on human imperfection in favor of artificial intelligence and the transhumanism mentioned above.

Brigitte Munier shows that science fiction has questioned through time the image of a man creating artificial creatures.

> (...) Science took over the Golem-robot: more and more works still question today the relationship between the human demiurge and the humanoid and super-powerful creature he made. [MUN 13, authors' translation]

Munier cites as examples *Metropolis*, *Blade Runner,* short stories, etc. Watching, for example, an episode of the Swedish series *Real Humans*, we see a possible society equipped with hubots (humanoid robots) whose objective is to assist us, to heal us, etc., revealing all the human physical and psychological weaknesses (lassitude, error, illness, etc.). But we also see them substituting themselves for human beings because of their supposed perfection, which is not without asking questions.

The last function is complementary to the reflective function. It is cognitive in nature and links the genre more strongly to a fictional status and narrative properties. Indeed, while science fiction is conducive to reflection it is because it enables an experience of thought, an immersion. Often dystopian, it places at the heart of its intrigue political, health or economic problems that are widely dealt with; moreover, technical objects are regularly the markers or indicators of these problems. This is how the spectators or readers can in return imagine solutions (technical or not) because they are driven by the narrative scheme that leads them to the solution of the problem. The fictional nature of science fiction, which is often far removed from reality, allows for greater freedom in experimentation [RUM 16, authors' translation]: "In its own way, by playing on the levers of the imaginary, science fiction offers cognitive resources likely to stimulate forms of hope and to open up experimental spaces for them," affirms the politician in this sense [RUM 16-17, authors' translation]. Fiction is sometimes more likely to allow us to have an internal vision of things, an access, a more obvious mediation:

> Thanks to the sensitive experience that it implements, literary fiction proposes a vision of the inside that is superior to the vision of the outside, a prisoner of conceptual thought. [TOF 17, authors' translation]

2.4. Science fiction and the imaginary world at the heart of training

Is science fiction in the teaching of human and social sciences in engineering schools truly mobilized? A quick consultation of the website of several schools does not really answer this question because the curricula are not always detailed. There is a good chance that artistic teachings oriented toward literature or cinema will draw from this pool. However, teaching entirely dedicated to this kind of education and to the development of a technical culture as we have defined it does not seem to be frequent.

Our experience at INSA Lyon[5], which supports our reflection/proposal, is part of this "configuration": while technical culture guides our contribution to teaching, science fiction is only one of the ways to achieve this[6]. The stakes are then high: it is a question of ensuring that a personal, free and entertaining culture becomes training material, an object of knowledge and reflection that can be integrated into future professional practice. This is not always self-evident because it requires bringing together cultures that are often opposed: literary culture and scientific culture [SNO 59]; both students and institutions are not always ready to take the plunge.

2.4.1. *Exploring science fiction representations*

This approach is rather that of an introduction to the question of representations, a question that will be explored in detail later. It is indeed a question of relying on a fiction of anticipation to give access to the representations conveyed by science fiction and to analyze the issues at stake. Here, the function of projecting science fiction stories into possible worlds is used. What are the future worlds envisioned? What does this tell us about our present world? Why? What questions can we take away from it? The first example is the reading in class of the prologue entitled "Un monde parfait" from Eric Sadin's critical book *La vie algorithmique* (2015). This essay begins with the story of a man's typical day, which is perfect and smooth because it is regulated by algorithms: the temperature of his bed, the nature of the tea depends on constants, the routes are optimized, etc. The subtle exaggeration of this text then allows us to question this possible world. During the session, each student reads aloud, in turn, a few lines from this book. Thus, as the reading progresses, everyone gets involved, imagines and projects himself into the world described. Do we want to live in this convenient, practical, efficient and perfect world? The collective discussion

5 The aim of this chapter is not to provide an exhaustive overview of all educational experiments incorporating the analysis of science fiction works in an engineering or science university course. Other experiments exist, no doubt more disruptive, and testify to the desire to adapt to new audiences, to favor logics of immersion and playfulness for which the imagination and in particular science fiction can play a central role. But we have taken the side here to deal with what we have experienced.

6 In order to give future engineers access to the understanding of user logics, we invite them, for example, to carry out field surveys, question practices, etc.

is built around the issues related to digital technology: that of permanent measurement and freedom, privacy, laziness and immortality.

The second example concerns robotics: it is based on the analysis of a sequence from the film *Ex machina* (Garland, 2014)[7] showing the particularly oppressive interaction of artificial intelligence with a computer scientist who is supposed to test it. During this sequence, the young computer scientist is asked to determine, using a kind of Turing test, whether artificial intelligence has self-awareness. The session aims not only to understand that through the director's choices, there are strong representations that feed a collective imagination around humanoid robots: she is a woman, she is made of metal, blue fluids pass through her, she has a pleasant plastic, nice facial features, etc. But this work also allows us to question a central issue in science fiction: that of the overtaking of human intelligence by that of machines. The scene shows a concrete illustration of the doubts expressed by human beings. It also highlights the power of these machines.

The use of science fiction has another virtue, which is in line with those mentioned above: by becoming aware of the technological imaginary world of works of fiction, engineering students are more willing to understand the projections and representations of users to better define and understand the relationship with technology from both an individual and a collective point of view. Indeed, if the student engineer plays the game to the end, they will have access to their own fears and the hopes they place on the technology, but they will also be able to carry out the same reflexive work by projecting themselves on those of potential users. Finally, they will also be able to wonder what disruptions will be caused if this or that technology is deployed. It is probably at this level that they will have access to the political aspect of technology. For example, they may ask themselves the question of free will: what does it mean for our freedom of thought? Or again the question of the selection of individuals: are we today in a eugenicist society? What does this mean about how we accept or reject disability?

7 The works that we work on (particularly movies or "television" works) are those that remain accessible to the target audience while at the same time presenting profound issues that are not treated superficially and are topical.

The pedagogical objective of these three examples is then to open up the discussion between the students and to make them question themselves while allowing them to ask the values they attach to the technology.

2.4.2. Science fiction to build an ethical approach

For a long time, science fiction has been thought of as a particularly fertile breeding ground for thinking about ethics. As we know, the stories it produces carry apocalyptic or prophetic imaginings that, if we are aware of it, encourage ethical reflection. Like Dominique Vinck, we believe that "the study of the imaginary world has a role to play in ethical reflection" [VIN 88, authors' translation].

Also, the first form of awareness explained above is discussed more in depth later on in terms of studies and in particular in a biosciences department by a module focused on ethics. The objective is to use fiction, and more specifically anticipatory fiction, to think about ethics and set up a real approach.

It is a question of moving away from a Manichean vision of technology or from the restrictive view of an ethics of good and evil. On the other hand, using the world of science fiction enables requestioning the "ethical imagination" [PIE 12] by creating new worlds into which to project themselves. Indeed, the narrative properties of science fiction allow engineering students to become aware of the ideologies associated with certain technologies [MIC 14]. For example, by analyzing certain science fiction films that deal with biotechnologies or medical technologies, engineering students question both the myth of perfect health [SFE 95] and the imperfection and vulnerability of humanity [BRE 13].

Also, the teaching module is divided into two parts. The first one is devoted to the analysis of cinematographic fictions (films or series), most of them coming from science fiction. The films analyzed are chosen by the students according to their sensitivity like that from *Gattaca* (Niccol, 1997) to *Westworld* (2016–2018) via *Terminator,* or the *Alien* saga. This first working phase allows the analysis of the representations of science and technology conveyed by this type of fiction. It is followed by a second one devoted to the construction of an ethical approach. For example, as the student engineers analyze the way robots are represented in these different movie narratives, they are questioning the nature of our humanity. What distinguishes us from a *Terminator,* a robot that is both destructive and protective at the same time, what can *a contrario* bring us closer to the robots of *Westworld*, those humanoids capable of consciousness and memory? By analyzing these representations, they can, for example, ask themselves about otherness, about this otherness which, in turn, allows us to know ourselves better. This stage of analysis is not insignificant because it lays the groundwork for individual and collective ethical reflection: what seems acceptable? What are the elements that can be discussed, negotiated and those that are not subject to any possible negotiation? These elements are then remobilized in a complementary work or when, to deepen their reflection, the engineering students conceive devices like an ethical matrix – on Ben Mepham's model – organize citizens meetings[8], or write argumentative texts such as the opinion of the *Comité consultatif national d'éthique* (French governmental advisory council on bioethics issues) on a given question. The construction of this ethical approach finally continues with their representation during a day where, confronted with other students, teachers and lecturers, the student-engineers present the results of their ethical reflections.

This work allows future engineers to look back: what was feared or what promises were made in the 1980s – or be forward looking – what science and technology promises us today and what are the apprehensions of our time and for the future.

This analytical approach is followed by what can be described as a more "productive" stage. Indeed, we ask the engineering students to produce a story themselves, either in the form of a fictional short story or in theatrical form. This stage is therefore more personal and allows us to take the work previously constructed around the representations further. Therefore, ethical reflection is not constructed *ex nihilo*.

Indeed, by writing, engineering students become aware of their own ethical position and defend the values that are important to them. Writing and, moreover, narratives help to structure thought, to put it in order and to

8 The ethical matrix is a tool to assist ethical reflection. In 1996, it was designed by Ben Mepham, a biologist and ethicist. One of its virtues is that it takes into account all pluralisms (values and actors) and does not limit ethics to an overhanging position. On the contrary, it enables ethical reflection to be integrated throughout the innovation process.

make scattered elements coherent. Thus, when the student engineer is placed in a writing position, the ethical experimentation function of the narrative is fully expressed. The fictional writing inspired by the analysis of these science fiction stories allows the student-engineer to forge an ethical position [KEM 97].

> Literary anticipation makes it possible to write what one wants – whether one desires or fears, it is basically the same thing – about what may happen. [KLE 16, authors' translation]

For example, during the writing workshop held in 2018, a student engineer staged an epistolary exchange of letters between emancipated artificial intelligence and a young man, locked up in a psychiatric hospital because he was in love with a virtual entity. Through this writing, the student composed a world where artificial intelligence is more human than humans in its ability to feel love and empathy. Among other things, he was therefore able to question his personal way of defining what humanity is and how the boundaries between man and machine are defined.

In the end, it is easy to see that the use of the imaginary world through science fiction operates as a fertile ground to conceive, think and even imagine ethics is quite convincing. Fiction makes it possible to go beyond, to free oneself from reality and to engage in an uncomplexed discussion where one does not worry about whether such a thing is possible, plausible or probable.

2.4.3. *Perspectives: harvesting and building on science fiction imaginary worlds in order to innovate*

During a broadcast on France Inter on August 10, 2018, Alain Damasio, the science fiction author, said that his books are not only there to describe dystopian societies and alert public opinion to possible abuses, but that they are also there to open up possibilities. As mentioned above, every technical object carries within it its functionality and its "fictionality" [MUS 14], and to try to separate these two elements would lead to a misunderstanding of what that object is. This "design assistance" function could be exploited in a training module on innovation in the specialization year. One of the aims of this would be to propose a reflexive stance on technology throughout the design process. This proposal would then make it possible not to think of technology only in terms of its impact on society, once the object is thought out, but on the contrary to integrate it from the design stage.

The aim of this approach would be to encourage student engineers to draw on the fertile soil of the imaginary world by drawing inspiration from the objects or universes they have seen represented or described in science fiction.

To do this, a training module could be set up by exploiting the imaginary world precisely in the design process. Indeed, partly based on the methodology developed by the MODIM Chair,[9] it would be interesting for the engineering students to become aware of the imaginary world of science fiction and use it to innovate. In their book *Innover avec et par les imaginaires*, written in the framework of this chair, Stéphanie Coiffier, Pierre Musso and Jean-François Lucas develop a mapping methodology to model the imaginary world in order to foster innovation. It is then a matter of identifying a "rationality of the imaginary world" [MUS 05, authors' translation] that can be formalized in the form of analysis grids. Therefore, based on the different steps they deploy – establishment of a spatially and temporally ordered corpus of documents, feedback on past experiences of use and dramatized stagings – develop a similar process to encourage student engineers to embark on an innovation process. They could conceive a mapping of the technical object's imaginary world they are working on, at the time of the ideation phase with the instruction to focus more specifically on science fiction stories and to draw up a heuristic map from the works, universes and objects of various technical fields. It is not only a question of accessing shared representations, but also of finding inspiration, of finding sources of motivation to improve an object or imagine a new one.

In the later phase of using creative methodologies, student engineers could integrate elements from science fiction using a tool like the discovery matrix such as lightsabers, spaceships, and flying cars and cross-reference them with the elements or functions of an object to be invented or improved. This passage through science fiction would encourage decentration, imaginative audacity and detours, before returning to a more realistic phase.

9 The MODIM Chair (*Modélisation des imaginaires, innovation et création* [Modelling of Imaginations, Innovation and Creation]) aims to develop fundamental and applied research around experimentation and training on the imaginations of innovation players and brings together researchers and industrialists between 2010 and 2015.

2.5. Conclusion

To conclude, we can say that while using science fiction works to access projections and beliefs about technology and innovation is not new, using it as one form of accessing an "all-encompassing" and non-segmented technical culture that can help future engineers to work in a technical democracy is perhaps a little more so.

We can recognize that this approach is not reserved for engineers and technicians as we are all concerned by technological development. Science fiction, because of its art of detouring, can also be a good material to launch citizens' debates in different mediation devices. This proposal could also be criticized for being difficult to transfer to the corporate world, and this is undoubtedly true, especially since science fiction is sometimes called upon for "bad reasons" [MIC 14]. Nevertheless, we hope to have convinced some future engineers that their work is part of society and that it is culturally appropriate. Finally, it seems to us that science fiction has three major assets.

It is a fictional genre that, because it amplifies, sometimes characterizes, gives very easy access to the meaning of technology through the representations and in particular to those of the technical objects that "populate" it. Its narrative power makes it a strong media lever: while "the utopia of science fiction is one of those micro-discourses that mediate technical progress in many sectors" [MIC 14, authors' translation], it also constitutes an entry point through which to begin to question the place of technology and find elements of an answer. It would then be a matter of turning these communication assets into a strength.

If we then move on to the field of innovation, we can also assume that the often offbeat nature of science fiction paves the way to awareness or even to innovations that break free from determinism [RUM 16]. Science fiction offers us a rich pretext to work with technology because it highlights the complexity of worlds and encourages us to think about it.

Finally, the critical and reflexive work made possible through science fiction could enable future engineers to have access to the meaning and significance of technology, to take their place in technical democracy and not to be locked into a technophobic or technoeuphoric Manichean vision, which is often paralyzing.

2.6. References

[BRE 13] BRETON P., "Pourquoi l'ordinateur n'a pas de parole?", in MUNIER B. (ed.), *Technocorps. La sociologie du corps à l'épreuve des nouvelles technologies*, François Bourin, Paris, pp. 87–102, 2013.

[CHA 94] CHANIAL P., BALANDIER G., "Entretien avec Georges Balandier", *Quaderni*, no. 23, pp. 119–132, 1994.

[CHO 11] CHOUTEAU M., NGUYEN C. (eds), *Mises en récit de la technique. Regards croisés*, Editions des Archives Contemporaines, Paris, 2011.

[CHO 15a] CHOUTEAU M., FAUCHEUX M., NGUYEN C., "Les élèves ingénieurs à l'épreuve de la fiction. Quelles relations entretiennent-ils avec les mondes scientifiques et techniques fictionnels?", *Les Enjeux de l'Information et de la Communication*, vol. 16/3A, pp. 69–82, 2015.

[CHO 15b] CHOUTEAU M., ESCUDIE M.-P., FOREST J. *et al.*, "L'ingénieur, au cœur de la démocratie technique", in LAMARD P., LEQUIN Y. (eds), *Eléments de démocratie technique*, Presses de l'UTBM, Belfort, pp.239–253, 2015.

[CHO 15c] CHOUTEAU M., ESCUDIE M.-P., FOREST J. *et al.*, "La technique est-elle condamnée à entrer par effraction dans notre culture", *Phronesis*, vol. 4, no. 2, pp. 5–16, 2015.

[CHO 17] CHOUTEAU M., FOREST J., NGUYEN C., "Quand la culture d'innovation fait écran à la culture technique", *Technologie et Innovation*, vol. 17-4, 2017. Available at: https://www.openscience.fr/Quand-la-culture-d-innovation-fait-ecran-a-la-culture-technique.

[CHO 18] CHOUTEAU M., FOREST J., NGUYEN C., "Concevoir en donnant du sens à l'innovation: l'approche P.S.I", in DIDIER J., BONNARDEL N. (eds), *Didactique de la conception*, Presses de l'UTBM, Belfort, pp. 73–87, 2018.

[CON 12] CONVERT B., DEMAILLY L., "Effets collatéraux de la création littéraire. L'exemple de la science-fiction", *Sociologie*, vol. 21, pp. 111–133, 2012.

[DE 90] DE CERTEAU M., *L'invention du quotidien. 1. Arts de faire*, Gallimard, Paris, 1990.

[DEF 96] DEFORGE Y., "Technique et culture", *Le nouvel éducateur*, https://www.icem-pedagogie-freinet.org/node/14116, January 1996.

[DE 81] DE NOBLET J., "Culture technique et changement de société", *Culture et Technique*, vol. 6, pp. 11–47, 1981.

[FLI 89] FLICHY P., "L'imaginaire collectif des ingénieurs: le cas des machines à communiquer", *Réseaux*, vol. 36, pp. 53–70, 1989.

[FLI 01] FLICHY P., *L'imaginaire d'Internet*, La Découverte, Paris, 2001.

[GAR 05] GARÇON A.-F., "Les techniques et l'imaginaire. Une question incontournable pour l'historien", *Hypothèses*, vol. 9, pp. 221–228, 2005.

[GUE 14] GUERMONPREZ P., *Les drônes débarquent*, FYP, Paris, 2014.

[JAC 93] JACOMY B., *Culture technique de l'ingénieur, Techniques de l'ingénieur*, 1993. Available at: https://www-techniques-ingenieur-fr.docelec.insa-lyon.fr/res/pdf/encyclopedia/tiagc-t40-version1.pdf.

[JAU 11] JAUREGUIBERRY F., PROULX S., *Usages et enjeux des technologies de la communication*, Eres, Toulouse, 2011.

[JOU 00] JOUET J., "Retour critique sur la sociologie des usages", *Réseaux*, vol. 100, pp. 487–521, 2000.

[KEM 97] KEMP P., *L'irremplaçable. Une éthique technologique*, Cerf, Paris, 1997.

[KLE 16] KLEIN G., "L'invention de l'avenir : prospective et science-fiction", *Futuribles*, vol. 413, pp. 29–52, 2016.

[LEC 96] LECOURT D., *Prométhée, Faust, Frankenstein: Fondements imaginaires de l'éthique,* Biblio Essai, Paris, 1996.

[MIC 14] MICHAUD T., "La dimension imaginaire de l'innovation. L'influence de la science-fiction sur la construction du cyberespace", *Innovation*, no. 44, pp. 213–233, 2014.

[MUN 13] MUNIER B. (ed.), "Le Golem ou les vertiges d'un homme fabriqué", *Technocorps. La sociologie du corps à l'épreuve des nouvelles technologies*, François Bourin, Paris, pp. 103–120, 2013.

[MUS 05] MUSSO P., PONTHOU L., SEUILLET E., *Fabriquer le futur. L'imaginaire au service de l'innovation*, Le village mondial, Paris, 2005.

[MUS 13] MUSSO P., "Le technocorps, symbole de la société technicienne", in MUNIER B. (ed.), *Technocorps. La sociologie du corps à l'épreuve des nouvelles technologies*, François Bourin, Paris, pp. 121–144, 2013.

[MUS 14] MUSSO P., COIFFIER S., LUCAS J.-F., *Innover avec et par les imaginaires*, Manucius, Paris, 2014.

[PER 98] PERRIAULT J., *La logique de l'usage. Essai sur les machines à communiquer*, L'Harmattan, Paris, 1998.

[PIE 12] PIERRON P., *Les puissances de l'imagination. Essai sur la fonction éthique de l'imagination*, Cerf, Paris, 2012.

[ROQ 83] ROQUEPLO P., *Penser la technique. Pour une démocratie concrète*, Le Seuil, Paris, 1983.

[RUM 16] RUMPALA Y., "Tester le futur", *Futuribles*, vol. 413, pp. 53–72, 2016.

[RUM 16-17] RUMPALA Y., "Entre imaginaire écotechnique et orientations utopiques. La Science-fiction comme espace et modalité de reconstruction utopique du devenir planétaire, *Quaderni*, vol. 92, pp. 97–117, 2016–2017.

[SFE 95] SFEZ L., *La santé parfaite. Critique d'une nouvelle utopie*, Le Seuil, Paris, 1995.

[SIM 12] SIMONDON G., *Du mode d'existence des objets techniques*, Ed. Aubier, Paris, 2012.

[SNO 59] SNOW C. P., *The Two Cultures*, Cambridge University Press, London, 1959.

[TOF 17] TOFFIN G., "La fabrique de l'imaginaire", *L'Homme*, vol. 221, pp. 167–190, 2017.

[VIN 88] VINCK D., "Imaginaires et biotechnologies", GEORGES T, PETER K. (eds), *Le triomphe des biotechnologies: la domestication de l'animal humain*, Presse Universitaire de Namur, Namur, pp. 211–217, 1988.

3

Engineers Versus Designers: Transposition of the Technical Imaginary World into the Visual

3.1. Introduction

We live in an era where the promotion of the "new" is one of the driving forces of consumer society. Innovative changes in the field of technology are dominated by engineers as the main agents – "masters of the economy" [MIC 14] of both tradition and modernization (at least since the rise of the Industrial Revolution). In this respect, engineering schools usually focus on technology per se and little on what the "beauty" of objects represents [CLA 09]. On the other hand, design, as a profession, is increasingly gaining visibility and influence. Nowadays designers take advantage of the fact that they are the ones who "put in shape" objects and therefore they have a great power to determine the direction of consumers' tastes. However, engineers have constantly tried to build their own visibility, while drawing public attention to the technology itself, to the mechanisms and to the work that is almost never visible but hidden behind the cases created by the designers.

We hypothesize that we live in an era that creates new forms of objects resulting from the readjustment of the limits of these two domains, or even a redefinition of their autonomy [BOU 96]. Supermodernity [GID 84] seems to be becoming an era that denounces art, fiction that deceives the eye and revalues everything that claims to be science by exploring an imaginary of technology.

Chapter written by Florin ALEXA-MORCOV.

The present study proposes to analyze the modalities of visual expression used more or less consciously by engineers in their attempt to become more visible while remaining autonomous in design and gaining prestige in the eyes of the modern consumer as promoters of the new technology, science, the new "magicians".

3.2. From applied science to applied art

Looking for "the new", one of the main expressions of innovation, is today disputed by two professional groups crystallized in time, that have never managed to remove a certain common zone of uncertainty – engineers and designers. "The new" becomes the main reason of identity construction and delimitation of these two groups, a subject of a power game.

While, until the 18th and 19th Centuries, the "creators" of material culture were particularly constituted by communities of trades or guilds, each oriented toward a certain field of activity and the production of a certain object, the nationalization of trades and professions radically changed the logic of the production of material culture [HIL 16]. The emphasis shifted from the finished object to its production in itself, both technically and socially. While, until the end of the 18th Century, the "creators" of material culture were particularly constituted by communities of trades or guilds, each oriented toward a certain field of activity and the production of a certain kind of object, the democratization of trades and professions, since the 19th Century, radically changed the logic of the production [HIL 16]. The emphasis shifted from the finished object to its production in itself, both technically and socially. Consequently, in our times what divide society is rather the components of the conceptual deconstruction of the object, a separation between form and function, between technology and aesthetics.

There is also an aspect that can be noticed from this confrontation of establish areas of action between these two professions: engineering is considered an applied science [HIL 16], engineers are technical people and base their activity on analytical thinking, while the design field is associated with applied arts, designers are closer to the arts and creative thinking is their main driving force. That was the starting point for a general public opinion that engineers deal with the technical part of a product and designers with its aesthetic part; engineers deal with the interior of the product, including the mechanism of that object (if it exists) and designers with the exterior of the product, with exterior appearance, with the social impact of that product. How did it come to this?

In fact, the dichotomy between "technology" and "aesthetics" is not recent, it is an old problem [LAS 11]. If the "technical" part of an object can be explained "rationally" and presented as such (sometimes insufficient to be socially accepted) in the case of the "aesthetic" part, the shapes of the objects, the relationship with the human factor and things become complicated. Many theories that have followed the "justification" of forms (natural or artificial) – mathematical formulas, "golden" proportions, etc. – have attempted a rationalist explanation of aesthetics and beauty. Architects and object creators (craftsmen, artists, designers. etc.) have tried to develop, in time, constructive or experimental solutions to justify the choices made in terms of shape, color, texture, proportion, etc., as a solution to persuade consumers that "beautiful" and "useful" could have a clear explanation. These include constructive solutions based on mathematical calculations (golden ratio, Fibonacci sequence, etc.); constructive solutions based on the study of the human body (ergonomics); constructive solutions based on the understanding and "copying" of natural principles (bionics); solutions for aerodynamic testing (using the already famous wind tunnels); solutions for testing the behavior of buildings subjected to external factors (winds, earthquakes, etc.); solutions for testing the behavior of buildings in the event of an earthquake, including the specialized software (this means a shift to virtual testing); and solutions for testing consumer behavior. Constructive solutions based on mathematical calculations such as the golden ratio, the Fibonacci sequence, etc., have also served as a basis for justifying a "good" proportionality between the elements of a whole (object or building). The dichotomy between "technical" and "aesthetic" has also manifested itself in terms of outstanding personalities in both engineering and design, each trying to demonstrate the importance of their field of activity in defining the "system of objects" [BAU 68]. The famous British engineer Dyson said:

> Like everyone else, we get frustrated by products that don't work properly. As design engineers, we do something about it. We're all about invention and improvement. (Dyson, www.dyson.com, accessed August 10, 2018)

Sometimes, even famous designers and architects have borrowed technical language and used it in their public speeches, one of the famous cases being that of Corbusier, who always tried to prove that his work was based on "scientific principles" and called, for example, kitchen furniture "equipment", a term used by engineers. Although, with the advent of industrial society, and in particular large infrastructure projects such as metal bridges, towers (the Eiffel Tower is one of the most relevant examples), etc., engineers gained visibility and prestige in society, yet, in terms of objects, designers were gradually favored because they were concerned with consumer behavior and their connection with the "system of objects". At the same time, the shift from the public interest of engineers to designers has been equated [LAT 12] with the idea of profitability and incentives for often unjustified consumption, as the external aspect of consumer products is considered essential in consumer choice.

> With the Great Depression, the movement (programmed obsolescence – n.a.) only accelerated, as the direction of American industry shifted from engineers to designers. [LAT 12]

One can justify Bourdieu's views that:

> In the heroic phase of the conquest of autonomy, the ethical rupture is always (...) a fundamental dimension of all aesthetic ruptures. [BOU 96]

The design process is promoted today as a global tool that has an influence on all social, economic and sometimes even political aspects, a real code that deserves to be deciphered in order to understand the evolution of a community or even of the entire society.

The "problem" for engineers is that they have almost always been "behind" their products and work, enjoying insufficient visibility, even when their work was related to stage activity, the case of theater engineers being eloquent [TKA 16]. Giulio Parigi's School of Stage Engineering, founded in Florence in the 17th Century, is one of the relevant examples of the involvement of engineers in artistic events related to theatrical performances or public, secular or religious processions. Both the Royal Courts and the Church needed magnificent, fabulous events in front of the public, made real by well-prepared "fictions" of "stories" (based on a whole series of special effects, as we call them today), and for this, "science" was needed, with engineers having an important role in this equation. This science of theatrical events engineering was so closely related that in the 16th Century, theater was considered to belong to the category of "mechanical arts" [TKA 16]. At the same time, designers gained visibility, sometimes becoming real stars, the most eloquent case being that of Philippe Starck, a French designer who promotes himself in eccentric situations, used rather by modern artists such as Salvador Dali or Picasso. He and his emblematic objects appear on the front page of magazines, and he has been invited to a whole series of conferences and to work with large industrial companies. On the other hand, engineers, such as James Dyson, have been presented in natural terms without eccentricities, close to his products, often showing their internal mechanism or simply how they work. This has been noticed not only by the new range of products that will "revolutionize" the functionality of certain household items, but also by a series of events: publication of books, involvement in the academic environment – creation of a Dyson University, a school created especially for engineers, etc.

Becoming more and more visible, also through public presentations as the already famous "design weeks", the work of designers is increasingly related to certain social and environmental aspects of public interest and in the same time their creative discourse is more fashionable than that of engineers. One of these creative approaches adopted in recent decades by designers is "green design" – one of the keywords of the last decades [WHI 93] – and its motto, "you are what you consume". Ethical issues related to the process of producing and obtaining raw materials have also been put into practice by movements such as "fair trade", initiated by The Body Shop, or "ethical consumption", taking into account aspects related to the production of objects from regions of the world where political regimes are despotic or where there are clear indications of human rights violations. All these events have made public opinion increasingly sensitive to this type of discourse and ready to punish any misconduct by manufacturing companies.

3.3. The question of the "object" in contemporary society

The Industrial Revolution marked a major change in the perception of objects. Baudrillard's very term "system of objects" clearly refers to a well-defined network in which each component (object) has a well-defined role, their functionality being closely linked to the role they play in a society [BAU 68]. The Industrial Revolution also implied the large-scale institutionalization of the mass production of objects (the assembly line being one of its most powerful symbols), seen as a process of democratization of material culture: everyone has the right to buy, to acquire the objects they want, each with identical technical characteristics, one of the myths of humanity becoming reality.

The concept of organizing production based on rational principles (promoted by design schools such as Bauhaus, 1919-1933) – including the economy of means of production; the use of "honest" materials strictly adapted to the product's purpose; an integrity of surface; a rational, non-sentimental, functional and serious design – led to a "logic of form" linked to the primary function of the objects and its relationships with the users [WHI 98]. The climax of this approach was the promotion of the "type-form" concept, expressing the conviction that there was an optimal solution for every functional problem, an ideal one; in this respect, consumers don't need to have a variety of products to choose from. The objects were to become aesthetic and timeless. The case of the rotary phone that appeared in the 1930s was an example, a standard, simple, mass-produced and impersonal form that has remained in the minds of consumers as a symbol of communication.

The design process emerged in modern society as a "mediator" between technical products, technology in general and society, which is too unprepared and too unwilling to accept the "new" in a short period of time. The whole process of adapting the machine to humans was difficult and focused on a first phase of camouflaging the mechanisms in the cases. Thus, the shape of the new complex objects (which usually included a mechanism) manages to define, in the minds of the users, the objects themselves. The shape of the object, the one that gives "meaning" to an object and gives it its own identity, is ultimately the one that distinguishes it from the other. In fact, designers tried to create an "affinity" between consumers and the new technical products that invaded the consumer market, and this whole process was also determined by a functional and semantic deficiency of the technical products created by engineers for consumers. Another important discovery about the way in which objects in society have evolved is related to their surface, namely that we now live in a world of smooth, shiny and bright objects, a trend that is also accentuated by new material processing technologies: chrome plating, polishing, lacquering, etc.; plastics, in turn, make an important contribution to this trend. Rough surfaces are now "interpreted" as unfinished, at an early stage of production or as belonging to a qualitatively lower range.

The physical need to put a whole series of mechanisms and machines in "boxes" in shapes, a global packaging of mechanisms, led to a new phenomenon: when they were technologically up to date, it was necessary to see this change visually, hence the appearance of the concept of style (design being understood as an added value). At the same time, the increasing speed of cars and trains led to the development of the concept of aerodynamics, which implicitly led to the creation of a new category of objects – the aerodynamic range – a real race in the struggle with the aerodynamic coefficient, a struggle that often led to purely aesthetic rather than functional solutions, as was the case with the automobile created by Norman-Timbs in 1948, which was never mass-produced due to its inadequacy in relation to the real needs of the consumer society. People's preference for round objects [HEL 11] rather than sharp-edged, right-angled objects, and a preference for safe handling, also led to the creation of a range of products with complex and spectacular shapes. For instance, designers such as Luigi Colani (1928-2019) develop spectacular biodynamic forms in order to change public perception about ordinary objects, such as vehicles or home appliances.

In contemporary society, an object-oriented society [KAH 15], the "object" as a symbol of material culture has gained increasing importance and visibility in the social sciences, being seen as a solution for understanding the functioning of the social world, anticipating its previous marginal condition. The "system of objects", increasingly physically attractive and more present in our lives, has become a "measure" of our society's development, a "document" [BER 16] that can be analyzed as such, with works increasingly devoted to developing a methodology for their study and the way we perceive their characteristics. As a result, a whole series of characteristics have been identified that influence our ability to understand and decipher the objects around us. One of the features of interest in this study is the difference between "the mental model" and "the conceptual model" [WEI 11]. The "mental model" represents those thoughts, feelings and information that a consumer has in their mind about how a certain product should be based on past similar experiences or knowledge of similar products, information from other consumers, or direct experience with such a product. People always have in mind such models, which is why user-centered design researchers have as their goal the very understanding of mental model of product's users. Conversely, the "conceptual model" is the way in which a product has been designed (thought of as a mode of operation and interaction) by designers, engineers, etc., the people involved in the process of creating and developing that product. If there is a discrepancy between these two models, problems arise: the product is difficult to understand, to use or not to be accepted by consumers. Changing a "mental model" of a new product to match the "conceptual model" may require training sessions, including advertisements that "teach" users how to use the product. It is advertising that has a major impact on people's mental models. We also talk about an association between objects and words. The ability to recognize objects is also closely related to our mental dictionary and the cultural context in which we live – if we do not recognize a particular object, we will try to associate it with something – it looks like ... – using verbal and visual metaphors. To better define the idea of a "mental model", it is worth recalling the term "affordances", which represent visible clues to the proper use of an object. These indications should be as clear as possible to consumers in order not to misuse the product concerned [WEI 11].

Studies on the complexity of objects [HEL 11] have shown that this can increase the attractiveness to consumers. Complicated objects are more attractive than simple objects, complexity being understood as the number of individual elements constituting an image or shape. Symmetry is another element that can increase interest in a particular shape.

As a result of all these discoveries, one of the important trends in the appearance of contemporary objects is related to the "dissolution" of their cases and the "visualization" of their functioning mechanisms. One of the relevant examples is metal wind-up toys, mechanically driven toys. For example, in contrast to the vintage versions, produced in the 1980s in China, new contemporary versions made by a design studio in the Netherlands (https://kikkerland.com/collections/wind-ups) have been released from the case, with the inner mechanism exposed. With the disappearance of the case, the decoration no longer has a place, so that the attractiveness of the object is given by the components of the mechanism and not by the strident colors that cover it. Note the red accents on the "insect's" feet to draw attention to them. These are made of plastic to protect the surfaces on which the toy will operate, but they also help to make it easier to slide on as many surfaces as possible. Both objects mimic species of the animal kingdom; in the first case, the reference is clear (including the movements made by the bird – pecking), and in the second case, we are dealing with an "insect" (a sensation given by the fact that insects are generally associated with animals that have "legs" first and move quickly, somewhat disordered in space). The second object certainly has an ambiguous, abstract and neutral aspect, closer to the contemporary period. The role of both solutions is to animate something lifeless, an older desire of people to create objects with autonomy of movement. The loss of the key would condemn the object to a static, immobile life. Both variants fall into the category of durable solutions – the fact that they do not "consume" batteries to be moved helps protect the environment, while being robust enough to leave a "legacy" for future generations.

3.4. The "transparency" of technology

Technology, often "hidden" behind "cases", is increasingly "exposed" and brought to the forefront by a range of conceptual approaches that have given rise to new series of objects, "forms" and means by which the general public is invited to discover them. It's also an invitation for consumers to discover the "engineering" behind the products and more than that, a way to reveal the "fiction" of inside mechanisms and technology. This process of "transparency" is ultimately a direct consequence of the change in the way the consumers understand the use of a product: increasingly pressing pollution problems; pressure to make public the way a product is made, the moral aspects; phobias related to automation and the excessive independence of objects in the system (Internet of Things); and so on. Communication in contemporary society is increasingly focused on the visual, the "image", the "highlighting" of technological progress that changes the material culture that surrounds us from one day to the next.

> The word 'imagination' definitely suggests that we can also think in images. Visual language is defined as a system of communication using visual elements. The term visual language in relation to vision describes the perception, comprehension, and production of visible signs. Just as people can verbalize their thinking, they can visualize it. Its structural units include line, shape, color, form, motion, texture, pattern, direction, orientation, scale, angle, space, and proportion. [KAH 15]

The work of designers is very much linked to this way of thinking "in images", sketches and drawings by hand or computer being a means of transposing thoughts into images. Engineers also use visuals, but here the drawing is closer to a Cartesian approach and technical drawing is preferred over artistic design. Mockups are also important elements used in the "visual discourse" of designers and architects, their purpose being to convey projects, solutions and ideas to the general public before they become reality.

3.5. "Transparent" objects

One of the most accessible solutions for object creators to show the "technology" to the general public is to replace the opaque cases designed to hide the operating mechanism with transparent covers that make it possible to observe the efficiency of a technical system that is more advanced than the previous one. One of the best examples is the Dyson vacuum system, which marked a major change in the cleaning products sector and subsequently developed other innovative technical systems. The transparency of the cases has contributed to the success of this new machine, as the suction process is visible to all.

We can also talk about transparency in the 2000 version of an Apple computer. The use of semiopaque polycarbonate has made it possible to create a product that is different from the traditional opaque and out-of-date cases used until then by all manufacturers. Although in this situation consumers do not really have anything to see inside, the transparency of the cases has staged a way of presenting the "new", a show of light and shadow designed to impress and demonstrate technological evolution. In addition, the new shapes had more rounded corners, the aerodynamic character being obvious and the "user-friendly" aspect being appreciated by the consumer public. The color variants have also contributed to a nice product, totally different from the gray and beige of the past.

3.6. "Deconstructed" objects

The technical presentation of objects, specific to engineers, inspired by the technical drawing, the "exploded" view, is used today as a new way of presenting new technical products. This way, the consumer is "helped" to understand the product's internal structure by using different colors and different types of representation. All of these are meant to help consumers to shape a "fair" mental model of the product and its use.

3.7. "Printed" objects

3D printers have also helped to "reveal" not only the technology hidden in some cases, but also the production process of the objects (true, for the moment, rudimentary). In contrast, the 2D printer process is hidden, partly because it is dangerous for the eyes and partly because there is not much to see. In the case of 3D printers, the process becomes part of the discourse of the object in question. Transparent protective glass, or the lack of it, exposes the entire process and turns it into a public spectacle.

3.8. "Skeleton" objects

The shift from analog to digital products (the case of many musical instruments) was another way of "exposing" the technological progress expressed by removing much of the volume of the cases from these objects. Even if the object often loses volume, the initial shape is usually suggested by a skeleton designed to convey to the consumer the category of object to which it belongs. Certain classic consumer items, such as watches, have given this name – skeleton – to a whole range of products promoted on the market under this name and whose mechanism is more or less exposed in order to reveal the complexity of the internal structure. Also, it is a new way of expressing the hard work behind the product.

3.9. "Impossible" objects

Fascination for technical objects and systems that defy natural laws and common sense have always been ways of attracting the consumer public. These paradigm shifts (Thomas Kuhn) completely change the public's perception of how an object or category of objects should be (and present itself), the case of the Segway being well known. In this case, the concept of dynamic stability is fully exploited, the "mirage" of two-wheel single-axle travel being in itself a means of impressing the consumer public.

3.10. Conclusion

The "staging" process (another expression related to the theater world) of a technical imagination has led to the creation of categories of objects and particular visual representations designed to "bring" consumers closer to new products, but also to the "magicians" found behind these products, who feel a strong need for recognition and appreciation, who have been left far too long in the wings – the engineers. The new objects and their representations, presented as a consequence of the innovation process, in the framework of "performative utopias" [MIC 14] are designed to make consumers aware, at least in part, of the hidden dimension of the innovation process. At the same time, however, designers, through their role as mediators between the technical and the social, strengthen their position as "creators" of material culture, being those who "put in shape" (i.e. "meaning") to objects, including those with a significant technical component. From this relative overlap of interest and action in the two fields, a new creative approach has emerged, which is linked to the "system of objects" that surrounds us.

3.11. References

[BAU 68] BAUDRILLARD J., *Le système des objets*, Gallimard, Saint-Amand (Cher), 1968.

[BER 16] BERNASCONI G., "L'objet comme document", *Artefact*. 2016. Available at: https://doi.org/10.4000/artefact.307, Accessed June 12, 2018.

[BOU 96] BOURDIEU P., *The Rules of Art: Genesis and Structure of the Literary Field*, translated by Susan Emanuel, First published in the U.S.A. by Stanford University Press, Stanford, CA, 1996.

[CLA 09] CLAY R., *Beautiful thing. An introduction to design*, Berg, Oxford, 2009.

[GID 84] GIDDENS A., *The Constitution of Society*, University of California Press, Berkeley, CA, 1984.

[HIL 16] HILAIRE-PEREZ L., VERNA C., "Histoire économique et histoire des techniques (XVe–XVIIIe siècles)", *Artefact*. 2016. Available at: https://doi.org/10.4000/artefact.292, Accessed February 16, 2018.

[KAH 15] KAHANE J., *The Form of Design. Deciphering the Language of Mass-Produced Objects*, BIS Publishers, Amsterdam, 2015.

[LAF 06] LAFFONT H., "Un ingénieur est-il bien *an engineer*? Contribution à l'analyse contrastive du milieu professionel de l'ingénieur en France, au Royaume-Uni et aux États-Unis", *ASp*. 2006. Available at: https://doi.org/10.4000/asp.766, Accessed September 30, 2016.

[LAS 11] LASKI G., Le design: Théorie esthétique de l'histoire industrielle, Université Paris-Est, Paris, 2011.

[LAT 12] LATOUCHE S., *Bon pour la casse. Essai sur l'obsolescence programmée*, Les Liens qui Liberent, Paris, 2012.

[MIC 14] MICHAUD T., "La dimension imaginaire de l'innovation: l'influence de la science-fiction sur la construction du cyberspace" *Innovations*, vol. 44, pp. 213–233, 2014.

[TKA 16] TKACZYK V., "L'École d'ingénierie scénique de Giulio Parigi (1608-1680)", *Artefact*. 2016. Available at: https://doi.org/10.4000/artefact.354, Accessed May 12, 2018.

[VER 98] VERIN H., "'Ingénieur': L'identité de 'l'ingénieur': quelques repères historiques", *Recherche & Formation*, vol. 29, pp. 11–20, 1998.

[WEI 11] WEINSCHENK S., *100 Things Every Designer Needs to Know about People*, New Readers, Berkeley, CA, 2011.

[WHI 93] WHITELEY N., *Design for Society*, Reaktion Books, London, 1993.

4

Imaginary Worlds to Be Projected or to Be Criticized? Methodological Considerations

4.1. Introduction

Marginal within managerial practices as recently as 5 years ago, design fiction – a speculative approach to design aimed at conceiving potential worlds in order, in a generally dystopian way, to provoke advanced reflection on its consequences for contemporary choices – is today employed and taught in many institutions, particularly in the context of strategic thinking. Little research, however, questions the effectiveness of this approach, whether it is its effects on a group in terms of projection, debate or the acquisition of resilient skills, to cite three benefits discussed in the specialized literature [AUG 13, COU 18, MIN 15].

In this chapter, we present reflections from ongoing research on the contribution of popular culture imaginary worlds to the activity of projecting credible visions of the future. Imaginary worlds are here defined as representations of a future in a form that can be figurative or narrative. Using the example of the foot soldier – richly represented in science fiction works (novels, comic strips, films) – we wish to explore different levers to exploit this material to generate credible proposals for the future.

The research perspective we present here is based on two hypotheses:

– the first is that mobilized imaginary worlds are by nature of diverse qualities, and that their projective capacities are just as diverse;

Chapter written by Nicolas MINVIELLE, Remy HEMEZ and Olivier WATHELET.

– the second methodological hypothesis that we formulate is that we can distinguish different levels of focus of the imaginary worlds due to the creative choices of their designers, which guides the work of appropriation. This variation in scale will focus on at least two levels that we propose to question: the use of "imaginary" technologies in laboratories to test their functioning, and the stage of changes in the sociotechnical environment that offer the possibility of creatively understanding the possible transformations of the world.

In this chapter, which is essentially focused on methodological issues, we will show how a large body of imaginary worlds can be understood at different scales. We will start by presenting the different technologies relating to the superficial aspects deployed in the literature, showing the existence of obvious imbalances in the forms of augmentation described. We will show that these are far from being fanciful, but respond to issues really studied by professionals in the sector. We will also see that the scenes described in the case of potential war situations – through the example of new vulnerabilities – do indeed point to real issues complementary to contemporary approaches.

The second part of our presentation will be devoted to the nature of the relevant fictional components to be engaged in a design-fiction approach: the context and the described uses. This last approach is certainly the most fruitful, as it offers the opportunity to test possible implications of interaction choices, an analysis that is very valuable within the design process. In this respect, "prediction errors" and interactions whose problematic nature "jumps out" to the viewer are also good thoughts for considering the development of a new technical solution.

Our aim is thus to explore how the imaginary worlds of science fiction can be used concretely by designers to reasonably speculate on the future of technical solutions. There are some indications that this speculation practice already exists, which we will discuss briefly at the end of this chapter.

4.2. Challenges in the production of a corpus of imagination

As part of our project, we created a database of nearly 300 imaginary images of the disembarked soldier. Four main mediums were solicited: films and cartoons; comics and manga; novels and short stories; and video games. With the exception of novels and short stories, whose diversity and volume of publication make them obvious resources, the other sources were chosen because of the visual character of the imaginary world. The participants in the workshops that we conducted during debriefing sessions believed that they enabled a rapid appropriation of a complex idea and made multiple projections possible. We propose to link this dual property to the stakes of our first hypothesis, i.e. the capacity to couple openness and coherence, inviting us to question the nature of the material in this equilibrium.

To create this corpus, we visited bookshops and specialized stores, and then we went further by consulting fan sites or specific encyclopedias. Many video games had their own wiki where fans detailed the capabilities of weapons, armor and equipment in detail[1]. It was the same for the animes and manga[2]. Superheroes from the Marvel or DC Comics universe also had analyses available on sites such as Comic Vine and encyclopedias detailing their powers, increases and weaknesses, making them more accessible. This work was based on strong serendipity.

It should be noted here that the use of databases did not enable access to levels of detail relevant to the collection of imaginary worlds when it came to visual resources, and the automatic processing of book data has yet to be built to generate efficient filters. This part of the human factor, and in particular expertise, remains, in our opinion, the best solution, albeit time consuming and, by nature, uncertain.

We then evaluated 12 criteria ranging from the type of cognitive assistance to the improvements offered, the ability of foot soldiers to work in groups and the environment in which they were expected to operate.

These quantitative data were then put into perspective by a qualitative analysis of the imaginary worlds, focusing in particular on how the designers themselves had envisaged their creations. This last point is sometimes highlighted very clearly in authors' biographies or in footnotes to famous works, such as the recent re-publication of *Ghost in the Shell* accompanied by the comments of its main author. Screenshots of films or scans of comic strips thus allowed us to highlight certain elements, and to confront them with the experience of reality of one of the co-authors, a lieutenant-colonel in the Army.

---

1 *Starcraft wiki* presenting the CMC Powered Combat Suit, for example.

2 *Super Robot Wars Wiki* featuring *Tekkaman Blade,* for example.

From these different approaches, we could draw four main methodological elements. First of all, the fact that among this mass of imaginary worlds, not all were of the same quality. However, many of them demonstrated a real ability to represent current issues in an extremely realistic manner. Second, and perhaps most interestingly, they also provided us with very good vision of tomorrow's challenges that we could test and experience. Finally, the question arose as to how to use this fertile ground for creativity in an intelligent way by distinguishing between context and use.

4.3. Imaginary worlds of various qualities

The first challenge of our approach was collecting the imaginary worlds as such, some of which proved to be richer than others. Comic books, for example, featured superheroes with magical powers that allowed them to fly or become invisible. While it is interesting to compile quantitatively these powers and to point out that flying is the most recurrent ability (see Tables 4.1 and 4.2), when we look at how the imaginary world proposes to make heroes fly, a "magic" explanation is not of much use for the real world.

Types of improvements	**Percentage of occurrences in the imaginary world**
Strength	43.55
Agility	20.43
Improvement of sight	19.35
Improvement of hearing	10.22
Improvement of speed and endurance	6.45

Table 4.1. *The five most enhanced physical abilities in the imaginary worlds of our corpus of 289 works of fiction*

Types of improvements	**Percentage of occurrences in the imaginary world**
Ability to fly	52.26
Ability to jump	24.52
Ability to be invisible	23.23

Table 4.2. *The three most evoked improvements in the imaginary worlds of our corpus of 289 works of fiction*

Conversely, video games or some manga and cartoons offer extremely accurate visions of how the capabilities offered to soldiers work. In *Call of Duty*, the ability to become invisible is close to what is currently seen in systems such as those developed by BAE. Furthermore, this ability is made very realistic by the fact that it uses a lot of battery power and therefore has a very limited useful life. Between these two extremes, we realize that some media offer more potential than others. Video games, by their desire for realism, offer a vision of the future that is radical but that tries to be realistic, and therefore is pseudo-justified scientifically. Comic books, on the other hand, take more liberties on the subject.

While the imaginary worlds are therefore very diverse in terms of richness, those that are relevant nevertheless offer extremely rich visions.

4.4. Representations that are often appropriable and exploratory

A first exploitation of our database focused on identifying the different ways in which soldiers were "augmented",[3] whether through technology or "physiology" (improvement of the soldier's physical and cognitive abilities).

Flying came at the top of the list of augmentations. It is a classic element of mobility projections, as evidenced by the imaginary worlds and projects of flying cars, almost as old as the cars themselves[4]. While individual flight for the soldier, a project dating back to the early 1950s, does[5] not seem feasible in the short term, the subject came up regularly, like the *Flyboard Air* prototypes of the French company Zapata Racing, which recently attracted the attention of the American DARPA and the French special forces.

The search for increased strength was also recurrent (via, in particular, high-performance exoskeletons), immediately followed by an increase in agility and vision. It is about compensating for humanity's physical and physiological limitations in combat. Let us just take the example of fatigue.

3 On the question of the augmented soldier, see, in particular, No. 32 of the journal *Inflexions*: "Le soldat augmenté?", 2016.

4 For example, two current flying car projects: the *Aeromobil 4.0 STOL* (https://www.aeromobil.com/aeromobil-4_0-stol/) and the *PAL-V Liberty* (https://www.pal-v.com/).

5 The Piasecki VZ-8 *Airjeep* was, for example, a prototype airfield developed by the US Army between 1957 and 1959.

A 20-day study of U.S. Army artillery units showed that those with 7 h of sleep per day were very close to 100% effective. On the other hand, those who slept for only 4 h lost 15% efficiency and even became dangerous [GRO 08]. Hence, the early interest in psychopharmacology: caffeine and cocaine at first, then amphetamines and modafinil (which can keep you awake for 40 h) to keep you awake. Note that 35 million methamphetamine tablets were allegedly used during the *Blitzkrieg* between April and June 1940 [RAS 08]. This approach – and the risks it entails – also appears regularly in the imagination world. The mercenaries in the American comic book *Nash* (1999) who regularly take drugs are shunned by civilian populations who fear their impulsiveness after injection.

Some of the imaginary worlds are therefore particularly interesting from a projective point of view, offering clear representations of the current and future challenges facing the landed soldier. If we look at the way camouflage and invisibility is used in videogames, some companies have been developing for several years [SHU 06] dielectric meta-surfaces that make their carriers (robots, drones, vehicles, even fighters) invisible at certain wavelengths (up to 70% absorption for radar waves, even visible light is concerned). Camouflage projects based on these discoveries are already relatively mature, such as the *Quantum Stealth* from the Canadian company Hyperstealth, prototypes of which were presented in 2016[6]. The same applies to the ability to fly, with a few prototypes from civil society being analyzed by the armed forces to assess their potential.

Ghost in the Shell, among other things, allows us to tackle the cyber theme. In 1989, the author, who is very well informed about military matters and technological developments, proposed a work that, in many respects, could be considered visionary, referring us to current conceptions of future war, and allowing us to question them. For example, in manga, cyber-control is used before, during or after a physical attack, but always in an intrinsically linked manner. Prior to a physical offensive, the role of a cyber-attack may be to confuse the opposing troops or even neutralize certain elements.

As the military is well aware, any battle should now be preceded by a struggle for superiority in electromagnetic and cyber space. This task is extremely difficult given the characteristics of this "smooth" environment

6 However, these materials are very difficult to manufacture and require a great command of nanotechnology.

(open and extensive architecture). During a physical attack, the hacking of sensors in an area where the major has to intervene allows him to divert the enemy fighters. Faced with a cyborg armed with a *Gatling* gunner, he can advance straight at him, having hacked his vision interface. After a physical attack, it is often a matter of the fighters taking control of data storage or a processing center or connecting to the enemy's machines to further infiltrate them. We see here the premises of a new form of exploitation of an offensive. Also highlighted here is the issue of network security if one or more machines are seized by opponents. Of course there are emergency erase buttons on the radios, but will soldiers still be able to use them? Is not it possible to recover data even after erasure?

4.5. New vulnerabilities

Some works of fiction are even more interesting because they invite us to look at innovations from the other side, pointing out their vulnerabilities and offering a detailed description of their uses. We are not talking about "flying" in general, but about discovering an exoskeleton in the middle of a battle, a particularly dangerous situation (see the running gag at the beginning of the film *The Edge of Tomorrow,* 2014). In our corpus, the most represented vulnerability consisted of destroying an enemy's "augmentations" by means of an electromagnetic pulse or an electrical overload (Table 4.3). A good example of this can be found in the film *Pacific Rim* (2013) where a Kaiju "electrocutes" a Jaeger and echoes current developments in non-nuclear electromagnetic weapons (e-bomb). The two other most mentioned vulnerabilities were, in order, power failure and hacking.

Type of weakness	Percentage of occurrence in the imaginary world
Electromagnetic attacks	14
Dependence on its energy source	14
Hacking	14
Slow movement	11
Exposure of vital parts	9

Table 4.3. *The 10 most evoked vulnerabilities in the imaginary worlds of our corpus from 289 works of fiction*

On this last point, the *Apple Seed* series (1985) and *Ghost in the Shell* (1989) were well ahead. Almost every fight in these two Japanese productions involves hacking. The heroine of *Ghost in the Shell* can physically connect to any computer system and "dive" into its depths. In *Ghost in the Shell: The New Movie* (2017), the view of an enemy sniper is hacked to prevent him from aiming correctly. This is perfectly in line with the concerns of the military which, when it comes to discussing the future operational environment and drawing on feedback from current conflicts (Ukraine, in particular), often believes that any combat will be preceded by "conquest" stage of the electromagnetic and cyber environment [HUB 00]. The film *Kill Command* (2016) offers a critical interpretation of the cyber risks linked to the use of foreign electronic components[7]. It shows an industrial subcontractor connecting to a sniper rifle, gaining access to the software and disconnecting it. The Pentagon's decision in May 2018 to suspend the acquisition and use of commercial UAVs because of operational security issues was fully in line with this framework [ATH 18].

As the two previous parts have detailed, the imaginary worlds manage to put current issues into perspective, while presenting those of tomorrow, offering visions of new uses, or new technologies, as well as their limitations. The question now arises as to how these issues can be approached more concretely, depending on the contexts or uses presented.

4.6. Context, a first point of entry for appropriating the imaginary worlds

By context, we mean here the ability of the imaginary world to report situations in a credible, open and, as such, usable way to question the current projections that are supposed to address them. In the field of landed combat, which we use here as an example, this theme can be illustrated by the problem of augmentation, whether physical or cognitive.

In the imaginary world, fighters who have experienced significant augmentations often tend to rely on them excessively and forget the fundamentals of combat. *Appleseed* also offers the vision of a human heroine, Deunan Knute, who has such experience that she is able to take over in her battles with cyborgs, especially through by subterfuge. In a scene from the film *Appleseed Ex Machina* (2007), she blinds hostage takers with a

7 A risk also present in [INS 15].

grenade, allowing her to save the captives. Also in the same series, in collaboration with her partner (who is a cyborg), she has to get rid of particularly dangerous "autonomous robot mines". She does this by combining bait, her partner coming toward her and long-range shots. We are therefore far from the caricatured vision of the supremacy of machines.

These imaginary worlds thus make it possible to question the concept of technological superiority. There is a recurrent criticism of the "overzealous cybernetic temptation". Characters who have undergone too many improvements often tend to rely on them too much and forget the fundamentals of combat. In the excellent *Ghost in the Shell* – which, as early as 1989, blurred the boundary between humans and machines, endowing the former with cybernetic assets and the latter with intelligence – the only totally human major character of this series, Togusa, manages to match in combat many of his augmented or even totally cyborg comrades. This resilience of the human being, with its specific capacities, takes us away from the caricatured vision of a total supremacy of machines as it has been conveyed by other works of fiction, such as *Terminator*.

This must make us wonder about the technological superiority of an army. It is often said that the spread of dual technologies reduces its operational superiority. However, it should be stressed that the accelerated spread of tactical know-how to and between irregular groups is an equally serious threat: "the most innovative element in recent years has been the expansion of the capacity to produce soldiers" [GOY 16, authors' translation]. Boko Haram, for example, has demonstrated its ability to coordinate relatively complex maneuvers including fire support and movement. The Battle of Mosul, despite a 20:1 balance of power, highlighted the tactical capabilities of ISIL (Islamic State of Iraq and the Levant), including a combination of intelligence, indirect strikes, countermobility and maneuver. Moreover, non-state adversaries have a permanent capacity for adaptation, enabling them to integrate available technologies more quickly into their modes of action. The case of homemade drones is symptomatic of this trend. With equal means, it is possible to be more effective with imagination (importance of organizational and doctrinal innovations).

Moreover, the imaginary world invites us to envisage an "analog" future for the fighter, less dependent on technology and less detectable because of its low electromagnetic "radiation". The soldier's future is undoubtedly also low-tech [COK 15]. Finally, the imaginary world can offer concrete cases to reflect on the distance between the virtual representation of war on or near the ground and its present and future reality. All too often[8], military equipment is considered outside the realm of close combat. However, this "stand-off" vision and high-tech warfare is probably a form of "bankruptcy"[9], because of its cost and also it is not enough to translate a military victory into a favorable political settlement (in the end, you have to go to the ground).

4.7. Uses, another point of entry for appropriating the imaginary worlds

Once the context of the imaginary worlds had been studied, the question arose of analyzing the uses they permitted. Here, the focus was on their concrete modalities, and sometimes even usability, since fictional scenarios are partly shaped by creative constraints.

An examination of the production of the works of fiction shows that all the choices in terms of the "design" of objects and interactions respond to very different logics. From the reuse of available scenery elements – such as the illuminated signs so characteristic of *Blade Runner*'s Los Angeles – to quotations and references – common in the field of flying vehicles – and even the magnification of features to enhance dramatic effects, there is no shortage of examples of interactions that escape the objectives of plausibility. Sometimes, as in the case of *Ghost in the Shell,* the author provides this type of commentary in the margins of an edition of the book. They enlighten us on the scriptwriting processes implemented and, above all, on the various resources mobilized to "build worlds" in a coherent manner.

Should we therefore limit ourselves to the collection of imaginary stories that have been elaborated under conditions of strong likelihood, in the manner of "hard science fiction" novels named in this way because of the intention of their authors to create stories that are as plausible as possible scientifically? We think not, since these imaginary worlds are not only elaborated with a view to testing the future.

---

8 This is the case, for example, in many videos presenting future military equipment. For example, see https://wwwyoutube.com/watch?v=TDZkN08rA0Q.

9 General M. Yakovleff, conference "Strategic Landpower for the XXIst Century", Eurosatory/Ifri, June 14, 2018 (available in video format on YouTube).

As an example, let us take a series of well-known interactions: the ones we find from time to time in the *Star Wars* saga: Han Solo in command of the firing post, trying to reach enemy ships.

To do this, he uses a screen with two targets that he tries to juxtapose. One understands, when this is the case, that he is in an ideal shooting situation, and that *a priori*, one of the squares represents him, with the other square representing the enemy. But then, if the onboard computer is able to fully analyze a shooting conduct, why not leave it to the machine?

The armor belonging to the protagonists of Richard Fox's *Iron Dragoons*, for example, allows this:

> Gideon tossed a rock into the air. Roland's hand opened and the stone landed in his palm 'Good. Your armor does most of the calculations. Your mind provides the impulse; the armor does the work', says the hero Gideon's trainer.

In the example of *Star Wars*, deliberately caricatural, we understand, on a narrative scale, the importance of having Han Solo waving in his shooting stance. The solution to the problem of usage is then rather obvious, as automated fire control systems already exist for the Close In Weapon Systems (CIWS) of many warships. In short, it is a narrative trick to make the scene more exciting. However, the functional usefulness of such interaction is questionable. Avoid being trapped by decoys not detectable by machines? Contribute to the psychological motivation of the pilot during very long distance journeys that are both tiring and monotonous, etc? Error – which is then only a question of point of view and design logic – becomes a creative stimulus here.

Can we really talk about failed predictions? No, rather the limits of certain projections, which show fairly quickly the strength of the use with regard to the technological aspects. However, fiction tends to project the audience into a world of technical plausibility, with less concern for the value of use. To look for aberrations of this kind is therefore to highlight the role of the everyday, essential in the life of objects and the use that will be made of them. In concrete terms, a number of imagined products are simply not viable because their actual use immediately disqualifies them.

Works of fiction sometimes seem to be aware of these limits, oscillating in the same sequence between highly technological extremes and others that are more "traditional" but no less effective.

By studying gaps of this type, and by systematically showing simpler ways of interacting in a context in the future, error becomes a very useful way to identify what seems essential and robust in its possible uses, a way of taming technology, especially when it is projected in discourses that are sometimes very emphatic about their ability to change the future.

The fictional characters have a wide range of weapons at their disposal, from lasers to mental manipulation. To be realistic, these equipment, sometimes fantastic, must describe in detail their interactions with humans. It is a very popular subject, particularly in terms of uses and physical and cognitive ergonomics. In a scene from *Kill Command*, a sniper with connected lenses sees what the optics of his rifle are shooting. This allows him to hide behind a low wall, place his weapon there, and shoot without having to expose his head.

In the film *Spiderman Homecoming* (2017), Spider-Man wears a suit that offers no less than 800 shooting options. To find his way around, a "head held high" vision offers the possibility of selecting the desired type of shot directly with his fingers. But the use of the interface is laborious because the number of possibilities offered is excessive and the hero must look at his hands to make the choice.

This counterexample invites the spectator to ask himself the question of what could be cleverly positioned on the hands or forearms. In a stressful situation, the intuitiveness of the system is imperative. However, this is not always the case with current equipment such as digitized systems: "It is their blatant lack of user-friendliness as well as their rigidity that most repels users, who are used to juggling with ease with civil digitization, the Internet, chatting, sending SMS messages or attachments" [JAC 13, authors' translation]. Ergonomics and usability are a permanent concern today in the defense industry. It must be said that several weapons programs have suffered from problems in this area. This is one of the priorities of the latest version of FELIN, 1.3 to restore agility to the landed soldier[10]. The future FELIN system will seek to better take into account the soldier's basic needs: communication, observation, shooting, finding one's bearings and protection. It should integrate augmented reality, "decamouflage"

10 The cumbersome nature of the mission preparation phase does not encourage its use in training. This version introduces the concept of a "mission kit" allowing the fighter's equipment to be adapted to the mission. This standard has been equipping the French forces for 2 years.

capabilities, indoor geolocation, lighter and probably more stealthy protective clothing, connected objects, advances in the field of energy, etc. [NOI 18].

Another key resource appears to be the management of the soldier's information and communication. The digitization of the battlefield, which has been in the works for several years, had previously been imagined and "tested" in fiction. Briaros, *Appleseed*'s cyborg, is thus equipped with "big ears" that allow it to hide behind a section of wall and let only one camera protrude. The generated video stream can then be shared with team members and used to guide shots. This type of approach is extremely well developed in video games such as *First Person Shooter*. It includes almost systematic and ultra-realistic maps of the environment, with the display of enemies and friends, as well as possible proposals for relocation.

Here, the value of some imaginary worlds lies mainly in their ability to draw attention to the limits of technology. While the use of holograms for communication is visually very attractive in fiction, its real applications show many limitations. In the video game *Batman Arkham Knight* (2015), the hero calls his partners by means of a projection on his wrist. Batman is therefore forced to interrupt his action and focus his visual field on his arm to speak. This results in a dangerous disconnection with his environment and the ongoing battle, illustrating, unwittingly, the risks of attention management related to the uses of augmented reality. Conversely, a technology that can represent an environment upstream of operations can be extremely useful in preparing missions. A tool of this type is presented in the film *Black Panther* (2018) where the hero, T'Challa, uses a table above which the convoy he is to attack materializes. He can turn around, interact with his bodyguard and even interact with virtual components by taking them in his hand. Technologies of this type, a kind of "sandbox 2.0", are envisaged to model the battlefield [BER 18].

These latter examples show a third type of issue: the soldier's interactions with his ecosystem. Video games and certain films thus make it possible to test new collaborations, particularly with drones or robots. In the film *Elysium* (2013), Agent Kruger, in an attempt to escape from a pursuer, drops a swarm of mini drones in capsules in his left hand and directs them with a laser pointer in his right hand. Once they hit their target, Kruger triggers their explosion using the right wrist interface. The details of this man–machine collaboration are very instructive, yet it is a key issue. It is indeed a question of effectively integrating soldiers, robots and artificial intelligence into a combat system. The desired gain is based on the complementary forces of man and machine. For example, automated systems increasingly exceed human capacity when it comes to codified activities with clear objectives. But algorithms have difficulty "deciding" in the face of data[11] uncertainty and ambiguity. Following the example of the use of robotic mules in the film *Spectral* (2016), the preferred scheme for envisaging the robot of the future is not, initially, total autonomy. It is his collaboration with the combatant – each protecting the other and "dividing the work" – that could bring real operational added value.

4.8. Conclusion

Through these different levels of analysis – identification of the technologies mobilized within a thematic set of imaginary scenarios; analysis of the risks made legible by the scenarios and finally description of the concrete interactions – we hope to have shown to what extent the imaginary is a particularly fertile source for thinking about concrete operational issues (in particular, in this chapter, we have dealt with the choices in the development of augmentation technologies; the balance between analog and digital in the soldier's equipment; the development of targeted interfaces; battlefield information management; and the ability to interact with other members of his company). They are not randomly generated, and as such, are a laboratory for imagining, testing and projecting future developments, which requires continuing along the path paved here.

There are some indications that, in the field of the army as in that of design in general, this circulation of ideas is already at work, with the entertainment and leisure economy relying on consultants from the sectors concerned to formulate plausible proposals, and the sectors in question mobilizing in an explicit or tacit manner – left to the designers and engineers and their own amateur fan practices – these resources to inspire or guide their reflections.

In the first case, several works have attempted to document the links between cinema, for example, and science [KIR 03, KIR 11]. Movies renowned for their "realism" and anticipation capacity, such as *2001:*

11 Joint Concept Note 1/18, "Human-Machine Teaming", Ministry of Defence, May 2018.

A Space Odyssey, were developed with the help of a large number of company consultants. The IBM teams were thus involved in defining the operation of the famous artificial intelligence. In a work devoted to the dangers of storytelling, Salmon [SAL 07] illustrates with numerous examples the collaboration between the American army and Hollywood authors in this inspiring work to create the most realistic simulators.

In the second case, as we have pointed out in the case of the manga *Ghost in the Shell,* designers and scriptwriters may have to go very far in their speculations about the likelihood of forms or interactions. As shown above, staging issues sometimes outweigh truthfulness – for example, when Masamune Shiro depicts torpedoes equipped with a jamming system, he does not draw this visual effect to avoid *making the drawing incomprehensible*, and so he only mentions it in footnotes. Elsewhere, on the other hand, we are captivated by the attention to detail and realism of certain images, elaborated on the basis of a thorough reflection on the verisimilitude of the scene. In one example among others, the author explains that: "Chroma is a humanoid remote-controlled terminal, so it does not need to close its eyes when it receives a shock or exhale when it throws a kick. She simply imitates her pilot's reactions and her movements mean nothing to her"[12]. Similarly, in the postface to the recent republication of *The Fourth Power* from the Argentinean cartoonist Juan Rimenez, Christophe Quillien asks himself the question "*cartoonist or engineer?*" so much the author pushes the technological detail in his work. The latter explains:

> I wanted to show an environment and scenery in which functional technology predominates, with a maximum amount of detail. I always enjoy drawing, and I spend a lot of time adding incidental details to my ships, even if they barely appear in a scene[13].

The sketches accompanying this re-publication show, among other things, how assumptions about the arrangement of vehicle engine parts explain the visible shape of their bodywork. Far from being aesthetic details, they are the expression of a prior reflection on the functional engineering of the machines.

---

12 Explanation in footnote to [SHI 01].

13 Rimenez J., *The Fourth Power,* Humanoids, 2000, p. 254.

This vigilance, exceptional in its level of detail, is not isolated in science fiction. Like Sid Mead[14], the famous designer of the flying car from the movie *Blade Runner,* among others, it is a question of thinking like a "designer" or an "engineer" who has to solve real problems. This willingness to speculate according to strong constraints – those imposed by the fictional world – without compromising a style that also knows how to be "futuristic" is a creative lever that has been able to generate outstanding achievements by their verisimilitude. Even more so when there is a real professional porosity, a film designer can intervene as a vehicle designer (Sid Mead had industrial design practice, and he also collaborated with Raymond Lowy, for example, on a Soviet car project) or, more simply, as a video game designer (e.g. Bradley Munkowitz game designer for Sony and screen interfaces for films such as *Tron* and *Oblivion*). More recent interviews with film designers converge in this direction[15].

These imaginary worlds are sometimes very advanced experiments in the design possibilities of objects and probable situations. However, the current tools for producing and exploiting imaginary world collections are largely left to those interested in these approaches. The absence of a formalized protocol goes hand in hand with the absence of criteria for operationalizing these approaches. In this chapter, we outlined out a few avenues of work to move in this direction. We did so first by proposing to distinguish the type of exploitable knowledge (context VS interaction); second, by developing reading grids elaborated on an analysis of failures; finally, by envisaging to exploit these imaginary worlds as narrative starting points and not as an "archaeology of the future" that would have to be discovered. The challenge is to go beyond a historical or aesthetic reading grid to take into account the impact of the situation of receiving work in a context of concrete innovation.

Therefore, we suggest that a relevant field of research is the identification of the criteria that make it possible to appropriate the imaginary world; in other words, the capacity to (1) generate new projections with regard to recurring themes within a community of actors; then (2) create new paradigms in the face of a given problem. Doing these two things while ensuring that non-expert participants are able to appropriate these imaginary worlds is a key issue.

---

14 Interview published in "Petite histoire du Spinner", *Rockyrama special edition "Blade Runner"*, 2017.

15 See, for example, the excellent collection of interviews on the blog http://www.inventinginteractive.com/category/future/.

The approach we have developed here is to use the imaginary world as a starting point for testing the potential of a design. However, not all are equally relevant. Sometimes, enclosed and linked to a particularly coherent world, the imaginary world is difficult to transfer to an environment similar to the present time and a credible vision of the near future. It is therefore difficult to manipulate. Conversely, elusive and based on aesthetic effects – like the evocations of flying vehicles on Coruscant (*Star Wars*, episode 1) – they cannot serve as a critical test for possible futures. Between the two, there is an enormous diversity of imaginary worlds that constitute relevant creative materials as a starting point for design or as resources for testing and critiquing proposals.

We believe that the capacity of the imaginary world to constitute a laboratory for projecting credible and relevant visions of the future lies in the tension between the focus proposed by the imaginary world to show itself as credible to its audience and what it offers us as a space for creativity (since it cannot take into account all the constraints of the real world). A laboratory whose full relevance is, we believe, demonstrated by the example of the foot soldiers we have proposed.

4.9. References

[ATH 18] ATHERTON K., "Pentagon suspends commercial drone purchases and use", *C4ISR.net*, June 13, 2018.

[AUG 13] AUGE J., "Speculative design: crafting the speculation", *Digital Creativity*, vol. 24, no. 1, pp. 11–35, 2013.

[BER 18] BERGOUNHOUX J., "Airbus modernise les briefings de l'armée de Terre", *L'usine digitale*, June 15, 2018.

[COK 15] COKER C., *Future War*, Polity, London, 2015.

[COU 18] COULTON P., LINDLEY J., COOPER R., *The Little Book of Design Fiction for the Internet of Things*, Petras, Lancaster, 2018.

[GOY 16] GOYA M., "Du bon usage du soldat augmenté", *Inflexions*, vol. 32, pp. 93–106, 2016.

[GRO 08] GROSSMAN D., *On Combat*, Back Bay Books, London, 2008

[HUB 00] HUBIN G., *Perspectives Tactiques*, Economica, Paris, 2000.

[JAC 13] JACOPS Y., "La NEB: limites et plus-values", *Doctrine tactique*, vol. 27, 2013.

[KIR 03] KIRBY D., "Science consultants, fictional films, and scientific practice", *Social Studies of Science*, vol. 33, no. 2, pp. 231–268, 2003.

[KIR 11] KIRBY D., *Lab Coats in Hollywood. Scientists Impact on Cinema, Cinema's Impact on Sciences and Technology*, MIT Press, Cambridge, 2011.

[MIN 15] MINVIELLE N., WATHELET O., MASSON A., *Jouer avec les futurs. Utilisez le design fiction pour faire pivoter votre entreprise*, Pearson, Paris, 2015.

[NOI 18] NOIZET T., "Que devrait être l'infanterie en 2038?", *Fantassins magazine*, vol. 40, pp. 105–114, 2018.

[RAS 08] RASMUSSEN N., *On Speed. The Many Lives of Amphetamine*, New York University Press, New York, 2008.

[SAL 07] SALMON C., *Storytelling. La machine à fabriquer des histoires et à formater les esprits*, La Découverte, Paris, 2007.

[SHI 01] SHIROW M., *Ghost In the Shell 2. Man Machine Interface*, Glénat, Grenoble, 2001.

[SHU 06] SHURING D., MOCK J.J., JUSTICE B.J. *et al.*, "Metamaterial electromagnetic cloak at microwave frequencies", *Science*, vol. 314, no. 5801, pp. 977–980, 2006.

[SIN 15] SINGER P., COLE A., *Ghost Fleet*, Eamon Dolan, New York, 2015.

5

Marsism, from Science Fiction to Ideology

5.1. Introduction

Among the many technological utopias that fuel global capitalism, the conquest of Mars and the creation of a civilization on that planet appear to be one of the most ambitious [CHI 92]. This planet is presented sometimes as a hell hostile to all forms of life, sometimes as a dream location for a few scientists, entrepreneurs and engineers looking for new frontiers to explore and domesticate. The space sector is one of the most innovative, generating significant returns on investment that are useful for the progress of human civilization. While the major space agencies have long had a monopoly on access to space, new economic models have been emerging since the 1990s. Called New Space, this topic is mainly defined by the interest of billionaires and entrepreneurs in exploiting space, through tourism, the exploitation of asteroids or the creation of extraterrestrial bases conceived as the premises of an expansion of humanity in the solar system. There is almost no doubt that the conquest of space will be at the center of the next industrial revolution. It is important to study the imaginary world that has been anticipating this process for more than a century. Indeed, what is the impact of science fiction on the construction of the strategic discourses of innovators, and on the imagination of engineers and scientists whose implicit function is to realize humanity's wildest dreams [FLI 01, FLI 95]? How does this technical imagination work? [HOT 13]. What is its role in the development of

---

Chapter written by Thomas MICHAUD.

technological ideologies, and is it reasonable to think that the utopian technologies of science fiction are intended to be realized by the actors of Research and Development (R&D)?

This chapter discusses the contribution of science fiction in the construction of the discourses on the conquest and colonization of the planet Mars [BOI 87]. In the 19th Century, accounts of a possible Martian civilization gave rise to much speculation and storytelling on the boundary between astronomy and an emerging science fiction. It was not until the 1960s and the Mariner 4 probe that the first photographs of this planet revealed an arid landscape, probably lifeless and hostile to humans, which astonished the scientific community. Although many dystopias have emerged in recent years, we will recall the influence of certain works of fiction on scientists, utopians and entrepreneurs motivated by the conquest of this planet. We will present some elements of Robert Zubrin's discourse, the scientist who created the Mars Society and published several books that helped popularize the idea of sending manned missions to Mars in the coming decades. His best-known book, *The Case for Mars*, and his novel *First Landing* have helped popularize the idea of sending a manned mission to Mars since the 1990s. Very much influenced by science fiction, and in particular by the idea of terraforming, the subject of a trilogy by Kim Stanley Robinson [ROB 99], he envisions the practical modalities of a sustainable implantation of humankind on Mars, and the different steps that would enable the creation of conditions for the habitability of this planet. His utopianism has largely contributed to the most positive discourse at a time when sending humans to Mars was not reasonable from a budgetary point of view, and above all technologically, as the development of spaceships still requires major innovations to ensure the smooth running of this type of mission. Were Zubrin's arguments taken seriously by the scientific community in the years 1990–2000, or was he an enlightened man, fascinated by visions of future science fiction that were far removed from the pragmatic considerations of investors [DIS 00]?

Second, the chapter will focus on the futuristic discourse of billionaire Elon Musk. Well known at the beginning of his career for having made his fortune in the information and communication technologies (ICTs) sector, he is currently developing numerous technological projects that could change the living conditions of humanity in the medium term. His main project is to send humans to Mars to develop a city of 1 million inhabitants. Musk's discourse is heavily influenced by science fiction and he does not hesitate to inject imaginary themes into his discourses. This has the effect of attracting many actors eager for technoscientific adventures and challenges. Musk talks, for example, about the dangers of artificial intelligence, about the creation of the hyperloop, a supersonic train and about neuroconnection, which until then belonged to certain cyberpunk movies and novels. He is also interested in autonomous cars, and many other themes that tend to be more than pure science fiction as discoveries and innovations appear promising to make even the wildest dreams come true [EAS 08].

5.2. The Mars Society's martian imaginary world

The Mars Society was founded in 1998 by Robert Zubrin. Its goal is to promote Mars missions to major agencies such as the National Aeronautics and Space Administration (NASA). Zubrin is an engineer known to have written influential articles presenting a plan to send humans to Mars. The Mars Direct plan has been developed in the book *The Case for Mars*. Since then, the Mars Society has opened branches in many countries, especially in Europe, maintaining international lobbying for the realization of the Martian dream. While many scientists have supported this initiative, science fiction writers such as Kim Stanley Robinson and James Cameron have also shown interest. Conquering Mars is presented as a strategic issue. Many members of the association are interested in science fiction, conceived as a means of expression of future representations, but also as a means of advertising the Martian cause [GEO 00, DUN 13].

On his blog,[1] Pierre Brisson, president of the Swiss section of the Mars Society, presents some science fiction works written by members of the association. One of his articles is entitled "Utopie, science fiction, fantasy; l'imagination est le moteur de notre action". It evokes the novel by Richard Heidmann [HEI 17], a polytechnician, graduate of the *Ecole Nationale Supérieure de l'Aéronautique* and former director at the *Société nationale d'étude et de construction de moteurs d'aviation* (SNECMA), one of the designers of Ariane's engines. It is 2143 and there is a warning called the *Alerte à Mars City*. A reporter goes to Mars, where there is a colony with 50,000 members. Pierre Brisson also evokes a novel by Jean Marc Salotti,

1 See https://blogs.letemps.ch/pierre-brisson/2017/08/15/utopie-dystopie-science-fiction-fantasy-limagination-est-le-moteur-de-notre-action/.

university professor in computer science at the *Ecole Nationale de Cognitique* at the *Institut Polytechnique de Bordeaux*, secretary of the working group of the *Académie Internationale d'Astronautique* in charge of taking stock of manned Martian missions and drawing up a list of recommendations for international space agencies, and also a member of the Board of Directors of the Association Planète Mars. This expert presents in *Genèse martienne* [SAL 16] the difficulties that the first explorers of this planet will probably encounter. Two types of Martian science fiction works can be distinguished. Some deal with the difficulties of exploration, and others with the functioning of societies and even a human civilization on Mars [DE 03].

Hard science fiction develops stories that are as realistic as possible. Robert Zubrin's *First Landing* [ZUB 01] is a fictional account of his plan to explore Mars. In the space of just a few years, the Mars Society has become a powerful, particularly influential lobby group, which seeks to convince major agencies or potential investors in search of the pioneering spirit that drives humanity and more particularly the American people. The writing of science fiction novels by some of its members shows the influence of this imaginary world in the scientific and engineering communities, particularly in the space sector [CH0 15].

At the Mars Society's 20th annual international convention in September 2017, a panel discussion featuring leading science fiction writers was held to define the role of science fiction in shaping the visions and strategies of the pioneers in the exploration and colonization of Mars. Gregory Benford was one of the guest authors. He had notably written the foreword to Robert Zubrin's novel [ZUB 01] and has written successful works of fiction such as *The Martian Race* [BEN 99]. David Brin was also a reference, as he is the author of several best-selling novels and a member of NASA's advisory board on the Innovative and Advanced Concepts Group. Jerry Pournelle and Larry Niven also participated in the debate.

Science fiction is an imaginary matrix very present in scientific debates and among decision-makers in the space sector [JOH 11]. Involving citizens in exploratory missions is a necessity for entrepreneurs looking for budgets to finance their projects. Using science fiction to communicate with the public is a well-known method, at least in the United States, since the *Star Trek* series supported and stimulated engineers and scientists on lunar exploration missions [WES 98, WES 00]. Science fiction is a powerful stimulant of the imagination, a discursive element so mobilizing that certain lobbies such as the Mars Society use it to participate in the planning of promising projects for the conquest and colonization of Mars. In terms of major space agencies, the interest in science fiction is obvious. The European Space Agency (ESA) has produced the Innovative Technologies From Science Fiction (ITSF) report, a study listing the major contributions of science fiction to revolutionary or future technical innovations in the technosciences useful for the conquest of space. NASA also regularly organizes science fiction competitions, notably around the Martian project.

Is the Martian imagination necessary or an obstacle to innovation, since the profusion of new references has still not been accompanied by a successful manned mission to Mars? Are not films and novels likely to frighten audiences, potentially shocked by certain stories? In the film *The Martian* (2015) for example, the hero is abandoned on a hostile planet after a huge storm, his teammates leaving the base to return to Earth having considered him dead. The critics unanimously praised the realism of the film, which benefited from the help of several knowledgeable scientists on subjects such as survival on Mars, agronomy and climatology. David Mimoun, a Mars specialist and professor at Supaero, said that "*The Martian* is an excellent advertisement for the engineering profession" in 2016. Science fiction can also have a negative effect on certain innovations by spreading a negative, uninviting, even frightening imagination. Other films feature hostile Martians, as in *Ghosts of Mars*, or very difficult survival conditions. The Martian astronaut is often a hothead, conscious of participating in a historic task and willing to risk his or her life to be among the first humans to set foot on Mars.

Robert Zubrin is convinced of the need to spread a positive Martian imaginary world in society. The financial and human cost of a manned mission requires the popularity and support of a large number of citizens. An agency that unwittingly commits large sums of money to an unpopular mission risks its reputation and a major scandal if the mission fails and one or more members of the ship die.

We can see changes in the technical imagination as scientific knowledge evolves. In the Martian sector, a large number of works of fiction have evoked Martians when no rational knowledge about this planet was truly available. The end of illusions led the imagination toward a reflection that was both imaginary and pragmatic. It was necessary to assess the conditions for carrying out an initial pioneering mission, as if to contribute to a technoscientific reflection requiring major advances in R&D. Other, rarer works describe a more distant future, in which a Martian base ended up generating a colony, or even a civilization. This great project excites the imagination, and is the basis for future political practices conditioned by major discoveries by scientists and engineers in the coming decades. It should be noted, however, that science fiction is rarely produced by decision-makers and politicians. Scientists, however, are great readers and producers of science fiction, which has become an element of dialogue and communication between inventors and decision-makers [MCC 97].

Thinking about the future of the Martian missions through lobbying is not an easy thing to do. Robert Zubrin's Mars Society has often encountered difficulties in convincing people of the need to quickly fund a pioneering mission. For the moment, the decision-makers have not decided to make the Martian dream come true. Zubrin, however, presented a detailed project for a manned mission to Mars in *The Case for Mars*, with figures and arguments based on knowledge in the years 1990–2000. His novel illustrates this plan, and the dangers and benefits to humanity of going to Mars. Since 2010, the conquest of Mars has taken a different path. As NASA's manned mission was out of date, private ambitions emerged, notably through a savvy entrepreneur, Elon Musk, a billionaire with a passion for science fiction and for whom nothing seems impossible. He was inspired by Mars Society's ideas in the early 2000s, before taking advantage of the collective momentum of some billionaires who had made their fortunes in the new economy.

5.3. Elon Musk, a utopian entrepreneurial spirit

Elon Musk is a high-profile entrepreneur who made his fortune by creating PayPal[2], and has since invested in several technology sectors such as automobiles and space launchers. His background in physics and business is accompanied by extensive reading, including science fiction. In various interviews, he has indicated that he particularly appreciated Isaac Asimov's *Foundation* series, because of the theory of historical cycles that explains the growth and decline of civilizations. This idea prompted him to consider colonizing the solar system in order to prevent the end of humanity. He

2 PayPal is a company created in 1998 by several founders including Elon Musk. It is an online payment service that was so successful that it was bought by eBay in 2002, making the fortune of its inventors. In 2017, PayPal held 80% of the electronic payments market share.

placed Asimov's novels in his Tesla car sent into space in 2018, giving a symbolic dimension to this initiative. Elon Musk also reports being influenced by Robert Heinlein's novel *The Moon is a Harsh Mistress*, published in 1966, during the preparation of the Apollo program. This work of fiction imagines the colonization of the Moon. Musk got the idea to create an extraterrestrial base, particularly useful to define his project of Martian civilization. Douglas Adams's book *The Hitchhiker's Guide to the Galaxy*, according to him, helped to manage an existential crisis experienced during adolescence. It would be an exaggeration to consider that science fiction has integrally guided the strategic vision of this American billionaire. He is indeed endowed with a great culture and is nourished by multiple texts, in different sectors, such as biographies, or scientific works.

Elon Musk is a visionary entrepreneur. He masters communication aimed at sharing his wildest dreams with as many people as possible. Announcing the forthcoming creation of a multiplanetary civilization belonged, before Musk, to the world of imagination, science fiction and technoscientific prospective [PRA 98]. Few players in the space sector spoke of the imminent arrival of technologies viable enough and of significant capital to send a manned mission to Mars. Musk announces that he will accomplish this feat, and he will create a base or even a colony on the Red Planet in record time. This chapter focuses on the impact of science fiction on the engineers and entrepreneurs involved in the exploration and conquest of Mars. One must also consult his many statements to see that he envisages science fiction scenarios to project into a hypothetical future the civilizational, organizational and societal consequences of technological innovations that he develops with teams fascinated by a charming managerial discourse. Science fiction has become a unifying discourse, both for the engineers working in Musk's various companies and for clients who, when buying products from Elon Musk, feel they are part of a collective dream whose cultural and imaginary roots are to be found in works of Asimov or Heinlein [WEI 95]. Musk plays with his image as a genius technophile, imbued with geek culture, to guide collective representations of the future in research communities, particularly in the United States. He cultivates this image to the point that many fans and media consider him the real Iron Man. This superhero is a billionaire who has developed many innovations such as a super-powerful armor that allows him to fight against the forces of evil.

Elon Musk has broadcast a science fiction discourse that is helping to create a speculative bubble around his activities. While he is not a Marvel superhero, he has become a “superentrepreneur”, a hero of a generation of engineers and innovators who dream of having his aura and financial power. Each generation cultivates ideal-typical representations of entrepreneurs. Think of Henry Ford or Bill Gates, for example, who have become true myths because of their revolutionary approach to business and management.

Nonetheless, the Martian dream has received much criticism, notably from Kim Stanley Robinson, who believes that Musk’s plan “looks like a big science fiction cliché from the 1920s, imagined by a boy trying to build a rocket to reach the moon from his backyard”. Musk is not the only billionaire investing in the conquest of space. Since the beginning of the 2010s, the New Space movement has seen hundreds of private actors take an interest in the exploitation of the cosmos, conceived as a new frontier that could be a source of considerable profit for the most audacious entrepreneurs. Jeff Bezos, founder of Amazon, is in line with Musk’s space discourse. Rather than Mars, he is interested in the colonization of the Moon. The Earth cannot continue to be home to heavy industry, he said, because it is too polluting and dangerous. In the coming decades, he wants to transfer these activities to space, and more precisely to the Moon. In March 2017, he even presented Donald Trump (President of the United States since 2017) with a project for a spacecraft to make deliveries to the Moon. A few months later, Trump announced his desire to send humans to the Moon to prepare a Martian mission, ensuring the conclusion of Bezos’ and Musk’s dreams, in a state perspective, NASA being, let us recall, at the origin of most of the space dreams of an American nation fascinated by science fiction.

The President of the French space agency, the *Centre National d’Études Spatiales* (CNES), Jean-Yves Le Gall, was interviewed by several journals at the end of 2017 to share his vision of future missions to Mars. In *Challenges,* he supported a “realistic” conception of the conquest of space, as opposed to the “science fiction projects” defended in particular by Elon Musk:

> Today, nobody believes in the Elon Musk show. His presentation last year in Mexico, in Guadalajara, was a Hollywood show. There were some great computer graphics, but no one believes that a ship is going to go to Mars with one hundred people on board, let alone a one-way trip. I totally refute the comparison that is being made with the settlement of

> America. When America was colonized, you had three weeks by boat, and if you survived, there were grasslands, turkeys, fresh water streams, you could build a house and settle down. You're going to Mars, there's no atmosphere, there's nothing. You depend solely on the ship that brought you here. And the story of going to extract minerals, water, it's not even futuristic, it's beyond science fiction. One must know how to maintain reason[3].

According to Jean-Yves Le Gall, it will take between 100 and 200 years to realize Elon Musk's vision, which he likens to a science fiction delirium. French realism contrasts with a science fiction conception of the exploration and conquest of the most influential American actors. This dichotomy is indicative of the differences in approach between these two cultures. Le Gall is even tempted to equate Musk's discourse with a form of irrationality, even though Musk plans to send a manned mission to Mars as early as 2024, only 7 years after Le Gall's criticism. The future will tell whether Musk's American science fiction approach will have triumphed, or whether French realism was right to promote a more cautious approach to an ambition probably less well diffused in the collective imagination. The pioneer mentality of the American people is also probably more receptive to science fiction concepts than French or even European society.

5.4. The technotype of the extraterrestrial base

Martian exploration has moved into a new era in recent years, with the emergence of new players, some of them private, who are considering the establishment of human colonies on this planet in order to create a new civilization, or even a new form of interplanetary business. Science fiction has long fascinated scientists and engineers from major space agencies such as NASA and ESA. Science fiction competitions on the theme of humanity's presence in space are regularly organized to involve citizens, especially the youngest, in space R&D. The latter is cutting-edge research, often the source of major innovations that have a significant impact on the daily lives of billions of people. Science fiction is a recurring element in applied research. While fundamental research, which is more theoretical, is difficult to relate

3 See https://www.challenges.fr/entreprise/aeronautique/conquete-de-mars-personne-ne-croit-au-show-hollywoodien-d-elon-musk_492171.

to this imaginary world, the work of investigating the future applications of scientific discoveries is, on the other hand, close to science fiction, even constituting a form of *institutional science fiction*. This refers to the practice of imagining future applications or processes in the framework of agencies or institutions, such as companies or state R&D centers. Science fiction is often an inherent subculture in some organizations in the most innovative sectors [MIN 16]. Discoveries in the space sector often seek to make science fiction come true. Traveling to the Moon means making the dreams of illustrious novelists such as Savinien de Cyrano de Bergerac, author of *The Other World: Comical History of the States and Empires of the Moon* (1657 and 1662) or Jules Verne, who published the prophetic *From the Earth to the Moon* in 1865.

The industrial revolution needs a technical imagination to feed the dynamics of capitalism, which is both its cause and consequence [MAR 95]. Science fiction is particularly active in times of high innovation, leading some people to fear that it could lead to disasters, if it were proven that R&D is making science fiction happen. Indeed, many stories depict dystopian worlds, or events in which human civilization is put to the test. The apocalyptic imaginary world of science fiction is a re-actualization of ancestral imaginary patterns, which have been ruling many societies for a long time. Technoscience is conceived both as the cause of disasters leading to the annihilation of civilization, and as the possibility for humanity to survive and control nature.

Moreover, science fiction is often set in the future. This specificity, which hitherto belonged to the oracles, is only rarely assumed by authors who do not wish to be presented as a new prophet form. Here again, we must go back a long way in history to have an explanation for such an intellectual stance [MIN 96]. For a long time, the predictions of the oracles were denounced because they did not come from God but from demons [CAZ 86]. It was therefore necessary to be wary of it and to prefer the discourses of the Church to predictions sometimes assimilated to a form of witchcraft, dangerous for the established order. Science fiction emerged after these religious considerations were subjected to revolutionary wrath, particularly in Europe. Thinking about the future became a new freedom, often appreciated by citizens, especially when it described societies in which progress made it possible to revolutionize lifestyles because of positivism [CHE 12]. Science fiction gradually became a very popular genre, making the masses dream, drawing on scientific discoveries to propose a new imaginary world, to a certain extent programmatic, for a scientific community increasingly erected as a dominant class, especially after the Second World War. Realizing the imaginary world became a leitmotiv leading to the advent of an era based on innovation and R&D [SFE 02, DRE 06].

Under these conditions, the conquest of Mars was part of this imaginary program, just like television, computers, the quest for immortality, etc. The extraterrestrial base, whether lunar or Martian, gradually emerged as a new imaginary world, as a technotype, that is to say a technological archetype revealed by innovation and technoscientific progress. Science fiction generates futuristic representations[4] of this technotype, feeding the scientific community with new projects and challenges. Since then, the imagination has taken hold of this technotype to propose multiple applications of this technical system recently made aware by humanity [DUR 93, GOL 95]. The first authors to tackle the exploration and conquest of Mars had only partial scientific data on the planet. Speculations about a possible Martian civilization were frequent and accepted in many learned societies. The first American missions sent to Mars showed that there was no civilization. Mars is a desert, probably hostile to humanity, and it would take considerable effort to make this planet a welcoming one. The Martian imaginary world then suffered from a pause. The new works demonstrated a new form of utopianism with the publication in the 1990s of Kim Stanley Robinson's trilogy (*Red Mars*, *Blue Mars*, *Green Mars*), which presented the terraforming of Mars as a new challenge for scientists. He envisioned the establishment of human colonies on Mars, with the aim of modifying its climate to make the planet habitable for humanity. Terraforming was a concept already present in the scientific community in the 1960s. But Robinson managed to turn his work into a new sectorial myth, enabling federating the scientific community around new futuristic representations [MOL 90]. The aim is no longer only to contact Martians, often feared, moreover, in the science fiction works of the 19th and early 20th Centuries. Research must now be directed toward new trajectories. It is not a question

4 The technovelgy site, which lists more than 2,500 technical inventions from science fiction works, indicates that the first mention of a lunar base dates back to 1930, in the novel *Brigands of the Moon,* by Ray Cummings. The *Moon Dome* formula was then used. The term *Moon Base* was first used in 1948, both by L. Ron Hubbard in *240,000 Straight Up Miles* and by Robert Heinlein in *Space Cadet.*

of preparing for a possible war against the Martians, but of learning how to manipulate the climate of a planet to make it habitable for humanity.

5.5. Marsism, nasaism, communism and technoscientific microideologies

The science fiction imaginary world is inspired by scientific discoveries that it confronts with societal problems often tinged with ideology [MAN 06, RIC 05]. The fusion of hard and social sciences generates this imaginary world, influential in scientific communities as well as in the lay population, yet indirectly contributing to the development of R&D policies and the funding of science. The meeting of science and democracy is facilitated by science fiction, a real interface between fiction and reality. The imaginary systems resulting from this process of extrapolation and exploration of possible technoscientific futures generate microideologies that subsequently feed the great technological ideologies, emerging from a form of underground. Before describing the transition from science fiction fantasy to *technoscientific microideology*, let us present a few examples of the latter.

Marsism is a technoscientific microideology insofar as the Martian imaginary world arouses interest in the research communities to the point of generating funding, and even guiding decision-making processes at the managerial and political level [SFE 95]. Initially, the Martian imaginary world was partly inspired by scientific discoveries. It also allowed them to go beyond their limits, a collective, popular, even democratic appropriation of its themes. Marsist microideology refers to the passage of the imaginary world into a new stage of collective reflection. The main works, some of which, such as Robinson's Martian trilogy, are sector-specific myths, are a horizon toward which many researchers and innovators tend to look. Terraforming Mars is not a short-term possibility. Nevertheless, the development of long-term mission plans is influenced by this science fiction. First, a film such as *The Martian,* critically acclaimed for its realism, is a model for future astronauts who will be sent on a mission to this planet. Science fiction generates stories that contribute to the education of engineers. It would be wrong to consider this contribution exaggerated, to the point of underestimating the major influence of scientific disciplines in their training. However, microideology plays an interface role between three major players in the decision-making and innovation process. The scientist, the decision-maker entrepreneur and the public assimilate science fiction discourses at various times, sometimes leading to the constitution of influential technological ideologies at a global scale.

Communicism is another example of technoscientific microideology that has been particularly active since the advent of the Internet. In this case, we have more hindsight than with Marsism, since the Internet has become a technological reality, which is not the case with the Martian base. While science fiction writers have since the mid-19th Century anticipated the main innovations in the field of television and networked computing, the cyberpunk movement, from the 1980s onwards, accompanied the development of the Internet. It would be a pity, however, to consider that science fiction did not predict the Internet, as most commentators claim. Cyberpunks and writers (1980–2020) were constantly imagining the future of the Internet as an immersive global simulation, also seen as an emerging technotype. While the ideology of communication described by Lucien Sfez [SFE 88, SFE 02] or the utopia of communication studied by Philippe Breton [BRE 00] allow us to understand the ideological and utopian phenomena underlying the emergence of this technology, we believe that science fiction has played a particular role in the construction of this ideology–utopia of communication. Breton devoted an entire passage of his book *Le culte d'Internet* to the study of Asimov's novel *The Naked Sun* [BRE 00] to illustrate the functioning of the utopia of communication. In the case of the Internet and the virtual world, the microideology known as communicism refers to the contribution of science fiction to the creation of a collective climate that facilitates, sometimes exaggeratedly or even deliriously, the financing of public or private R&D policies in this sector. In this case, science fiction allows the adhesion of a large number of actors who work collectively to the advent and interpretation of the immersive global simulation technotype.

Nasaism is a third neologism that designates a technoscientific microideology, this time institutional [PAR 92]. Indeed, NASA produces a large number of scientific discourses and avoids confusing the public, sometimes restoring reality when irrational discourses try to mislead citizens with false information. NASA plays the role of a regulator of the technoscientific imaginary world and sometimes organizes science fiction competitions to stimulate the imaginations of millions of space conquerors around the world. This type of structure, like the European Space Agency

(ESA) with *Esaism*, proposes what Lucien Sfez calls in *La Santé Parfaite* "fictions sciences" [SFE 95] whose function is to ensure a form of common sense with regard to scientific knowledge and to propose explanations on the origin and future of the human species in a very long cosmic history. NASA's discoveries have fuelled many fantasies and science fiction stories since its inception in 1958. It can also be seen as the consequence of a science fiction explosion in the United States for several decades [LUO 13].

In what ways does the science fiction imaginary world become a technoscientific microideology [MAL 00]? Initially, the imaginary world appears in authors with variable and protean references. Rather than speaking of a collective unconscious common to all authors and all humans, at this initial level, it seems that imaginary patterns appear in isolated individuals, although living in a culture that is both specific and global. In a second phase, some of these stories fall into oblivion, others reach a wider audience, while the most accomplished are elevated to the status of a sectorial myth. We are thinking about William Gibson's cyberspace and Kim Stanley Robinson's terraforming. At this scale, a selection is made from the sum of the stories, even if all of them contribute to proposing utopian technologies, imaginary manifestations of technotypes that innovators are responsible for raising awareness and materializing in the form of new marketable technologies. We have been able to analyze in previous publications how the virtual imaginary world gave rise to communicism, a microideology that eventually transformed into a particularly influential technological ideology. In the case of the Martian science fiction imaginary world, we have observed its transformation in the form of a microideology, Marsism, due to the appropriation by scientific and economic actors of its themes. This phenomenon partly explains the belief that science fiction is endowed with precognitive power. In fact, these fictions contribute to innovation as they are inserted into ideological discourses coming from actors increasingly close to political and economic decision-making processes. A virtuous circle is observed with regard to the influence of science fiction on decision-makers. Initially, a large number of them have been immersed in a culture that has been very much imbued with this imagination for several decades. In particular, science fiction on television has contributed to spreading a form of technological utopianism throughout the world, mainly originating in the United States. In a second phase, they continued to create a science fiction that they considered important in guiding R&D at a global scale. Gradually, managing and innovating became synonymous with the realization of an increasingly shaped and ideological technical imaginary world. Numerous institutional works of fiction by companies or space agencies such as NASA or ESA explained to the general public that imagination was important for innovation and that science fiction could be a source of inspiration, but also a particularly mobilizing mode of communication and storytelling. When NASA and ESA present their programs, their videos are on the boundary between science fiction and documentary. Numerous films on corporate foresight follow the same approach. Science fiction aesthetics has become a reference point for the actors in charge of planning the future and of convincing citizen consumers of the quality of the utopian society they wish to achieve.

In the space sector, new players are emerging every day. As New Space enthusiasts, they want to liberalize access to space and compete with technocratic organizations such as NASA. They have often adhered to an imaginary science fiction world, some of them dating back several decades. Colonization is a primordial instinct of the human species. It is therefore obvious that some actors, sometimes called "astropreneurs", devote part of their fortune to be pioneers of the ultimate frontier and pass on to posterity. Bases on extra-terrestrial territories are technotypes that science fiction has helped to map and represent. The New Space phase promises the transition from Marsism, hitherto used and promoted by lobbies such as the Mars Society or large agencies such as NASA or ESA to another dimension. New Space futurists are waiting for a technological revolution, particularly in the propulsion and engine sector, for Marsism to become a major technological ideology. The exploitation of certain asteroids or possible metals on Mars could generate a new type of economy and interplanetary capitalism.

5.6. Conclusion

The history of the exploration of the planet Mars has demonstrated the influence of scientific discoveries in the technopolitical imaginary world. In the 19th Century, it was thought that a Martian civilization had built networks of canals, even leaving some observers worried about possible hostility from these little-known neighbors. Science fiction helped to spread this idea [ALK 10]. Indeed, it became a popular fantasy in the United States during the interwar period. A scientific error by the astronomer Schiaparelli created a collective imaginary world that abounded, arousing the desire to better understand these extra-terrestrial people. Science was later at the origin of the discovery of the non-existence of Martians. For several decades, science was the source of collective irrationality. Science fiction was particularly rich at that time, testifying to an imaginary world generated by an erroneous theory. The enormous progress made in our knowledge of this planet has provoked a change in imagination, which is now directing humanity toward its interplanetary expansion. At the same time, imagination became an important element in the development of R&D policies and strategies. Capitalism no longer fears a possible war against technologically superior Martians. From now on, it is inspired by science fiction, presented as a source of technological utopias useful for the management of organizations and society [SEG 05].

5.7. References

[ADA 19] ADAMS D., *Hitchhiker's Guide to the Galaxy Trilogy*, Picador Books, London, 2019.

[ALK 10] ALKON P.K., *Origins of Futuristic Fiction*, University of Georgia Press, Athens, 2010.

[BEN 99] BENFORD G., *The Martian Race,* Aspect, New York, 1999.

[BOI 87] BOIA L., *L'Exploration imaginaire de l'espace*, La Découverte, Paris, 1987.

[BRE 00] BRETON P., *Le culte de l'Internet, une menace pour le lien social?*, La Découverte, Paris, 2000.

[CAZ 86] CAZES B., *Histoire des futurs, les figures de l'avenir de Saint-Augustin au XXIème siècle*, Seghers, Paris, 1986.

[CHE 12] CHENG J., *Astounding Wonder: Imagining Science and Science Fiction in Interwar America*, University of Pennsylvania Press, Philadelphia, 2012.

[CHI 92] CHIRPAZ F., *Raison et déraison dans l'utopie*, L'Harmattan, Paris, 1992.

[CHO 15] CHOUTEAU M., FAUCHEUX M., NGUYEN C., "Les élèves ingénieurs à l'épreuve de la fiction. Quelles relations entretiennent-ils avec les mondes scientifiques et techniques fictionnels?", *Les Enjeux de l'information et de la communication*, 2015, available at: http://lesenjeux.u-grenoble3.fr/.

[CYR 16] CYRANO DE BERGERAC, *The Other World: The Comical History of the States and Empires of the Word of the Moon*, Didcot House, London, 2016.

[DE 03] DE WITT D.K., *Astrofuturism: Science, Race and Visions of Utopia in Space*, University of Pennsylvania Press, Philadelphia, 2003.

[DIS 00] DISCH T., The *Dreams Our Stuff Is Made of: How Science Fiction Conquered the World,* Free Press, New York, 2000.

[DRE 06] DREGNI E., DREGNI J., *Follies of Science – 20th Century Visions of Our Fantastic Future*, Speck Press, Denver, 2006.

[DUN 13] DUNNE A., RABY F., *Speculative Everything: Design, Fiction, and Social Dreaming*, MIT Press, London, 2013.

[DUR 93] DURAND G., *Les structures anthropologiques de l'imaginaire*, Dunod, Paris, 1993.

[EAS 08] EASTON L., SCHROEDER R., *The Influence of Imagination: Essays on Science Fiction and Fantasy as Agents of Social Change*, McFarland & Company, Jefferson, 2008.

[FLI 95] FLICHY P., *L'innovation technique: Récents développements en sciences sociales, vers une nouvelle théorie de l'innovation*, La Découverte, Paris, 1995.

[FLI 01] FLICHY P., *L'imaginaire d'Internet*, La Découverte, Paris, 2001.

[GEO 00] GEOFF K., *Science Fiction Cinema: From Outerspace to Cyberspace*, Wallflower, London, 2000.

[GOL 95] GOLDEN K.L., *Science Fiction, Myth, and Jungian Psychology*, The E. Mellen Press, Lewiston, 1995.

[HEI 17] HEIDMANN R., *Alerte à Mars City – La colonie Martienne comme si vous y étiez*, Editions Auteurs d'Aujourd'hui, Vendres, 2017.

[HOT 13] HOTTOIS G., *Généalogies philosophique, politique et imaginaire de la technoscience*, Vrin, Paris, 2013.

[JAM 07] JAMESON F., *Archéologie du futur, un désir nommé utopie*, Max Milo, Paris, 2007.

[JAM 08] JAMESON F., *Archéologie du futur, penser la science-fiction, Tome 2*, Max Milo, Paris, 2008.

[JOH 11] JOHNSON B.D., *Science Fiction for Prototyping: Designing the Future with Science Fiction*, Morgan & Claypool Publishers, London, 2011.

[LUO 13] LUOKKALA B.B., *Exploring Science through Science Fiction*, Springer Science & Business Media, 2013.

[MAL 00] MALRIEU P., *La construction des imaginaires*, L'Harmattan, Paris, 2000.

[MAN 06] MANNHEIM K., *Idéologie et utopie*, Ed. de la Maison des sciences de l'homme, Paris, 2006.

[MAR 95] MARCUS G.E., *Technoscientific Imaginaries*, University of Chicago Press, Chicago, IL, 1995.

[MCC 97] MCCURDY H., *Space and the American Imagination*, Smithsonian Institution Press, Washington, DC, 1997.

[MIN 96] MINOIS G., *Histoire de l'avenir, des prophètes à la prospective*, Fayard, Paris, 1996.

[MIN 16] MINVIELLE N., WATHELET O., MASSON A., *Jouer avec les futurs: utilisez le design fiction pour faire pivoter votre entreprise*, Pearson Education, Paris, 2016.

[MOL 90] MOLES A., "La fonction des mythes dynamiques dans la construction de l'imaginaire social", *Cahiers de l'imaginaire,* vol. 5/6, pp. 9–33, 1990.

[PAR 92] PARKER M. (ed.), *Utopia and Organization*, Blackwell, Oxford, 1992.

[PRA 98] PRANTZOS N., *Voyages dans le futur, l'aventure cosmique de l'humanité*, Le Pommier, Paris, 1998.

[RIC 05] RICOEUR P., *L'idéologie et l'utopie*, Le Seuil, Paris, 2005.

[ROB 90] ROBINSON K.S., *Red Mars, Green Mars, Blue Mars*, Bantam Spectra, New York, 1990.

[SAL 16] SALOTTI JM., *Genèse martienne Tome 1 Objectif Mars*, Amalthée, Nantes, 2016.

[SEG 05] SEGAL H.P., *Technological Utopianism in American Culture*, Syracuse University Press, New York, 2005.

[SFE 88] SFEZ L., *Critique de la Communication*, Le Seuil, Paris, 1988.

[SFE 95] SFEZ L., *La santé parfaite, critique d'une nouvelle utopie*, Le Seuil, Paris, 1995.

[SFE 02] SFEZ L., *Technique et idéologie, un enjeu de pouvoir*, Le Seuil, Paris, 2002.

[VER 11] VERNE J., *From the Earth to the Moon & Around the Moon*, Wordsworth Editions, London, 2011.

[WEI 95] WEICK K.E., *Sensemaking in Organizations*, Sage Publications, Thousand Oaks, CA, 1995.

[WEI 14] WEIR A., *The Martian*, Broadway Books, Portland, 2014.

[WES 98] WESTFAHL G., *The Mechanics of Wonder: The Creation of the Idea of Science Fiction*, Liverpool University Press, Westport, 1998.

[WES 00] WESTFAHL G., *Space and Beyond: The Frontier Theme in Science Fiction*, Greenwood Publishing Group, 2000.

[ZUB 01] ZUBRIN R., *First Landing*, Ace, New York, 2001.

[ZUB 11] ZUBRIN R., *The Case for Mars: The Plan to Settle the Red Planet and Why We Must*, Free Press, New York, 2011.

6

Quo Vadis Engineering? Science Fiction as a Means to Expand the Epistemic Boundaries of Technoscientific Innovation

6.1. Introduction

This chapter looks at the "identity crisis" of engineering professions – theorized in 2002 by MIT's technology historian Rosalind Williams [WIL 02] as the "expansive disintegration" of engineering – as well as at the untapped epistemic potential of science fiction to expand what might be called: *the engineering mindset*.

In order to understand the so-called "identity crisis", we investigate the dual figure of technological inevitability that emerges at the confluence of two conspicuously distinct social imaginaries: on the one hand, the "convergence" imaginary [ROC 03, NSF 14] and, on the other hand, the "Fourth Industrial Revolution" imaginary (as per Klaus Schwab and the vast network of the World Economic Forum). As we move on to show, it may well be that today's institutional discourses of engineering professions lie at their intersection. Indeed, everything occurs as though the engineering mindset and engineer character (such as that embodied in *The Engineer of 2020*) act as an invisible symbolic mediator of contradictions lying at the crossroads of these two social imaginaries: i.e. two largely fictitious, but nonetheless operating figures of technological inevitability. The question

Chapter written by Marie-Luc ARPIN, Corinne GENDRON, Nicolas MERVEILLE and Jean-Pierre REVÉRET.

therefore arises as to whether and how the two figures of inevitability may be transcended.

In what immediately follows, science fiction literature is thus construed as a means for these two figures' mutual reinforcement, but also (and above all) as a potential means to overcome them. In the third part of this chapter, we then move on to define the two types of technological inevitability as converging toward, and emerging from engineering institutional discourses. As we shall see, their mutual alignment *and* contradictions may as such be conceived as expressions of engineering's "identity crisis". In the fourth part, we explore the mechanisms of what may be called the "instrumentalization of the social" through an engineering mindset; these mechanisms allow us to better understand how the "identity crisis" of contemporary engineering professions might well have been managed, up until now, to the detriment of epistemological pluralism: a pluralism which engineering institutions paradoxically claim to promote, namely by emphasizing inter- and trans-disciplinarity in engineering education programs and reform efforts. In the fifth part of this chapter, science fiction is finally construed as a means to overcome technological inevitability and to reconnect with the kind of epistemological pluralism so fundamental to genuine interdisciplinary endeavors.

6.2. Science fiction at the heart of engineering innovation

People often talk about how science fiction inspires scientific vocations and technological innovations. The *Star Trek* series continues to inspire scientists, engineers and astronauts,[1] while Nichelle Nichols' incarnation of Lieutenant Uhura is said to have inspired Mae Jemison, the first African-American astronaut[2]. But science fiction also feeds directly on scientific discoveries, and many science fiction writers are either researchers or have developed an impressive scientific culture. The formalization of the link between science and science fiction is not simple, however, because the parameters of each genre are distinct and even antagonistic, as illustrated by the exercise undertaken by Chromatiques in the interactive work 2101 where scientists were invited to comment on different episodes of fiction.

1 See https://www.nasa.gov/feature/50-years-of-nasa-and-star-trek-connections.

2 See https://today.duke.edu/2013/10/maejemison.

> It's very difficult for us, says Isabelle Vauglin, Astrophysicist at the Observatoire de Lyon. Patrick Chiuzzi urged us to imagine what *will be* possible, but it is not yet possible. Which amounts to asking *why* this is not possible at the moment: are the barriers technical? Theoretical? Can these barriers change? Aeronautics was unthinkable at the time of Marie Curie, it is not certain that a researcher of the time would have argued that it would be as much a part of our lives 100 years later. We were sort of heading down the same road.[3]

It is true, therefore, that even when absolved by the injunction to stretch the imagination the researcher may hesitate to engage in speculation, fearing that it will undermine their own credibility. Nevertheless, today's possibility often is yesterday's impossibility, whether in terms of travels, communication or even life domestication. In this sense, through the imaginary depictions it fosters, science fiction allegedly is at the heart of scientific innovation: it shifts boundaries and embodies new possibilities that can stimulate and even feed into scientific development.

Yet strictly construing of science fiction as a process aimed at responding to the injunction of innovation reinforces but a narrow perspective of technoscientific development. If indeed science fiction has the potential to alert us to the excesses of technological developments, at the very same time, it also tends to portray us as powerless in the face of science's great and inevitable march forward. From this point of view, science fiction appears to be subservient to the surrepticious paradigm of technological determinism or the underlying assumption of technology's inescapable autonomy.

Moreover, this paradigm is at the heart of tensions underlying the aforementionned engineering mindset, whose fundamental concept of problem-solving – commonly conceived as scientific and efficient – has spread to all spheres of human life over the last half of the 20th Century. Through this diffusion, a mode or model of governance has gradually imposed itself, whereby all spheres of human activity are bent to such a problem-solving logic. The same therefore applies to the so-called "grand challenges" (or great "social issues" of our time) that the engineering professions are now seeking to resolve through mostly *technical* means [NAT 08], while at the same time ignoring such means as challenges *in and*

3 See http://sciencespourtous.univ-lyon1.fr/science-fiction-pousse-chercheurs-dans-leurs-limites/.

of themselves or, more generally speaking, as impediments to the inverse logic whereby social integration might well transcend technoscientific development.

An alternative interpretation of science fiction, however, could provide fertile ground for broadening the engineering mindset, along with its narrow perception and understanding of social matters. Such an alternative – we argue – could be nourished by an anthropological and social interpretation of works of fiction, while at the same time giving way to works of sociopolitical anticipation. Departing from the foregoing problem-solving scheme and underlying engineering mindset, these works of sociopolitical anticipation would thus portray society first and foremost as the result of politics and social choices; they would also provide a measure of the sociopolitical determinants of technoscientific orientations (inevitably rooted in social struggles). This would thus presuppose a reversal of the current trend that locks the social and politics into a rationalizing apparatus, while offering a way out of what Edward T. Layton has called: "the engineer's dilemma"[4] [LAY 86].

6.3. Figures of inevitability: the engineer at the confluence of discourses

In January 2016, a 46th World Economic Forum (WEF) was held in Davos on the theme of *Mastering the Fourth Industrial Revolution*. On the foundations of the Third Industrial Revolution, this fourth revolution would be marked by a fusion of technologies that waters down the boundaries between the material, digital and biological. Its "disruptive" potential would surpass that independently conferred on Big Data, 3D printing, artificial intelligence, the Internet of Things or even the industry 4.0 that integrates them, so that it would imply nothing less than a transformation of the human species [SCH 16]. By fundamentally altering the way we live, work and relate, this industrial revolution would present promises, but also perils that no other time has known [SCH 16].

4 As Layton originally wrote in 1971, "The essence of the engineer's dilemma is, at base, bureaucracy, not capitalism. The engineer's problem has centered on a conflict between professional independence and bureaucratic loyalty, rather than between workmanlike and predatory instincts. Engineers are unlikely to become revolutionaries because such a role would violate the elitist premises of professionalism and because revolution would not eliminate the underlying source of difficulty" [LAY 86].

In the light of this proclaimed revolution, the world leaders gathered in Davos had the ambition to rethink our economic, social and politics systems in order to "shape a future that benefits all" [SCH 16]. This is how the discourse of the Fourth Industrial Revolution is constructed, which enshrines the *inevitability* of technological change while insisting on our capacity to shape it: "the world is fast changing, hyper-connected, ever more complex and becoming more fragmented but we can still shape our future in a way that benefits all. The window of opportunity for doing so is now" [SCH 16].

6.3.1. *The disruption-less discourses of disruption*

Carried and disseminated by the World Economic Forum network, the discourse of the Fourth Industrial Revolution intersects with and complements those of the circular economy, the Third Industrial Revolution or prosumption in an ideology of "disruption-without-disruption". The challenge is to reclaim a present in crisis for and by shaping the future. But reappropriation and shaping are based on the contribution of technologies that are both *controllable* and yet inevitable. Like a call to break with a present which they themselves help to perpetuate, these discourses place at the heart of their narrative the critical issue of the future of humanity, while making the end of current crises (environmental, social, etc.) dependent on a conception of technological change that exacerbates them in the present. So what happens, then, to other forms of change or, more specifically, those that are not primarily technological? The paradox in these discourses becomes clear: all of them (logically) assert that "technology" (or technological change) cannot be autonomous, and yet it has to be for them to thrive!

6.3.2. *The "convergence" discourse*

Technological change's inevitable character is exacerbated by the hypothesis and project of "convergence", which dominates the discourses of transdisciplinarity [KLE 14] and allegedly marks the distinction between the Fourth Industrial Revolution and previous ones. According to Klaus Schwab, this "convergence" is how "waves of further breakthroughs [are occurring simultaneously] in areas ranging from gene sequencing to nanotechnology, from renewables to quantum computing" [SCH 16]. Thus, these waves of discovery are conceived as the mark of a new era, but the research modalities behind them remain invisible and unchallenged. So that "convergence" only appears as a watermark in the discourse, through the evocation of transformative forces (a constant technological change) whose current and future implications we hope to master, without paying heed to its history or roots. In this sense, convergence is the black box of the Fourth Industrial Revolution: it is taken in and naturalized as the hallmark of autonomous technological change experienced as inevitable, rather than as a new research strategy for knowledge production.

Convergence, however, corresponds to a set of activities that are situated, collective, distributed and in the making: in other words, as Jarzabkowski [JAR 05] would put it, convergence is convergence-as-practice. So for those who practice it, there is nothing immutable or natural about it. It even conceals a certain vulnerability. Convergence is first and foremost a "transdisciplinary" approach to solving scientific and societal problems [KLE 14], which is based on a fusion of technologies and disciplines that are touted for their potential for "disruptive innovation" [NAS 14][5]. Like nanotechnology, which in itself constitutes one of its important axes, as well as one of the industrial sectors that crystallize it, technological convergence is deemed to be the only problem-solving approach capable of solving the unprecedented challenges of industrial and postindustrial societies by renewing the way we do things: through them, "all problems will in principle not be solved [...], but they will become null and void, obsolete. We will simply do things differently, in a radically different way" [DUP 04, author's translation]. In the present context of environmental urgency, for example, their development is so conceived and portrayed as *inevitable* by its promoters and participants.

The discourse of convergence thus cultivates a notion of *inevitability* that echoes that of the "disruption-less discourses of disruption". But the *inevitability* of convergence is one of necessity and purpose, not of principle

5 "Convergence is an approach to problem solving that cuts across disciplinary boundaries. It integrates knowledge, tools, and ways of thinking from life and health sciences, physical, mathematical, and computational sciences, engineering disciplines, and beyond to form a comprehensive synthetic framework for tackling scientific and societal challenges that exist at the interfaces of multiple fields. By merging these diverse areas of expertise in a network of partnerships, convergence stimulates innovation from basic science discovery to translational application. It provides fertile ground for new collaborations that engage stakeholders and partners not only from academia, but also from national laboratories, industry, clinical settings, and funding bodies" [NAT 14, p. 1].

or nature. The *inevitability* of technological change is therefore not an implicit presupposition: it follows from a chain of arguments whereby even the politics is mobilized. As probably the most influential movement in transdisciplinary research [KLE 14], convergence runs counter to institutionalized scientific research. Therefore, its *inevitability* ensues from an agenda rooted in the politics of technoscience, rather than in a sense of history.

6.3.3. *The engineer character at the confluence of discourses*

If the two *inevitabilities* nevertheless demonstrate confluence – we argue – it is because they both find an echo in engineering's institutional discourses. Through the discourse of leading engineering institutions – such as those of the National Academy of Engineering (NAE) and the Massachusetts Institute of Technology (MIT) in the United States – a dual vision of technology is constructed, and the two *inevitabilities* come to coexist, sometimes superimposed or merged, sometimes colliding with each other.

But this confluence comes at the price of an "instrumentalization" of the social: that is, a process whereby the institutional discourses of convergence become themselves *instruments* that are developed, acquire materiality and spread by their conflation with social theories. So although the engineer "being" admittedly gains in reflexiveness through such institutional discourses, it is far from certain that such a concept of transdisciplinarity paves the way toward an authentic epistemological pluralism. All the opposite, it can even be argued (as we do here) that these new transdisciplinary practices tend to have a closing effect, which is reflected by the notion of *instrumentalizing* the social.

Over the past few years, a series of reports have indeed been produced to illustrate this new (transdisciplinary) era of engineering, one of which has enjoyed particular visibility and influence since its publication in 2004 by the NAE [KER 13], namely *The Engineer of 2020: Visions of Engineering in the New Century*. In resonance with it, more recent narratives have emerged from engineering institutions, including the so-called "international movement"[6] of the *Grand Challenges for Engineering* [see, e.g., NAT 08, MOT 16, KOH 18, MOT 18, GRA 18]. Against such a discursive backdrop, the now widely accepted issue of the *future of humanity* is conflating with that of the *future of engineering* – undoubtedly more surprising and less moving. Through them, reformed engineering institutions are construed whose present and future are glorious, but that obviate the tensions underlying engineering's current "identity crisis". Such tensions constitute the background upon which the "instrumentalization" of the social by the engineering mindset is occurring i.e., the lens through which it must be understood i.e., and the process by which epistemic truncation and colliding effects emerge while compromising social creativity and imagination.

6.4. Instrumentalizing the social

The instrumentalization of the social, akin to the concept of rationality engineering, refers to the process by which "instruments" are developed, materialized and disseminated that tacitly incorporate derivatives of social scientific theories in the form of presuppositions or analytical categories:

> Rationality engineering refers to the process whereby tools incorporating rational choice theory's assumptions are developed. This process describes the material embeddedness of organizational decision-making processes within rationality. [CAB 11]

As such, the recent practice of "instrumentalizing" the social ties to an older practice, which according to French historian Héléne Vérin is at the basis of modern engineering, namely, "problem-forming" (translated from: "*mise en forme-problème*"). This process consists of circumscribing a field of potential solutions through the definition of a "problem to be solved" [VAR 93, authors' translation]. As a pattern of thought, the "problem-form" that ensues newly makes it possible to break with determinisms (or fatality) so as to aspire to transforming the world, or to carry out actions that influence its course, within a certain domain of *possibilities/solutions*:

> [The engineer's project] takes the form of a problem, that is to say, by definition, that which contains in the statement of its terms the conditions of possibilities for its resolution. Its very

6 See on this subject: [MOT 16, KOH 18].

> structure determines the range of possible 'demands', of possible choices, the validity of which it allows to control. [VER 93, authors' translation]

The "instrumentalization" of the social, so conceived, evokes a new era in which an engineer's mindset and action is born and directed toward issues whose technical nature is inextricably tied to social and politics agency [see, e.g., WIL 02, SHE 08, RIV 16, FRA 17], whereas this technical nature often implicitly roots in the appropriation and translation (in part or whole) of social theories[7].

Through *The Engineer of 2020*, in particular, the "instrumentalization" of the social is done through the appropriation and translation of *strategic forecasting* theory. This report confronts the question of the future of engineering head-on: in the context of a society that "continually changes", whereby engineering is required to "adapt to remain relevant" [NAT 2004], "changes in the engineering profession and engineering education have followed changes in technology and society" [NAT 04]. For *The Engineer of 2020*, which is embodied by this institutional discourse, the central issue was whether it still would be possible that "a future be created where engineering has a broadly recognized image that celebrates the exciting roles that engineering and engineers play in addressing societal and technical challenges?" [NAT 04]. It is as though this forecasted engineer felt invisible, or idle even, to such an extent that he/she "aspire[d] to a public that [would] understand and appreciate the profound impact of the profession on sociocultural systems […] [as well as] recognize the union of professionalism, technical knowledge, social and historical awareness, and traditions that serve to make engineers competent to address the world's complex and changing challenges" [NAT 04]. Against the backdrop of a technological, digital, "information" society, which nevertheless seems to be building its character as central, *The Engineer of 2020* is marked by a strange ambiguity. As if torn between a glorious past that all are presumed to remember and bleak prospects, this engineer is trying to overcome in the future an ontological confusion vividly felt in the present.

7 See, for example, [WIS 12]. Among other considerations, this author's historical analysis recounts how, in the United States of the 1960s, Jacques Ellul's theory of "technological society" was apparently appropriated, translated and disseminated by a group of engineers in search of social relevance and identity markers.

6.4.1. *"The art of the long view", or the theory of strategic foresight*

In order to guide the reflection that would result in *The Engineer of 2020* report, the NAE called upon Peter Schwartz: a renowned futurist and strategic planning consultant. Schwartz is well known in the American management community, and is the author of *The Art of the Long View,* originally published in 1991. In the 1990s, together with Kees Van der Heijden, he strongly contributed to the dissemination of scenario-based planning within the corporate world, a strategic planning theory which is now widespread [VAR 10]. Indeed, within today's management sciences, *strategic scenario planning* – also known as *strategic foresight* – has become a research field in its own right [BRA 05, VAR 10].

In his 1991 book, Schwartz insists particularly on two closely related principles: (1) *openness* to as many possible scenarios as possible and (2) *non-investment* in the scenarios. Under the principle of *openness*, the scriptwriting exercise should not be understood as a prediction exercise, because "it is simply not possible to predict the future with certainty" [SCH 91]. According to the "art of the long view", it would nevertheless be possible to foresee it "with uncertainty", so that ultimately one would be favorably disposed toward any eventuality, however improbable it might be or appear. Unlike prediction, which excludes and selects certain possibilities from the outset according to their probability, the "art of the long view" therefore aims very specifically to foster cognitive and emotional openness toward the widest possible (if not complete) range of future scenarios. In order to reach such a level of openness, and although Schwartz [SCH 91] does not state it in these precise terms, the scriptwriting process must in principle embody what is today conceived as a transdisciplinary process of knowledge production: through such a process, not only would the traditional boundaries of academic research be transcended to include the participation of non-academic actors, but also, a series of more fundamental boundaries such as the tension between cause and effect or fact and value would simultaneously be transcended [CAR 10].

In accordance with the corollary principle of *non-investment*, Schwartz also stresses the importance of remaining vigilant in the face of the strong propensity of most people to fix attention, once the scenarios have been generated. Thus, rather than encouraging individuals to take hold of the knowledge produced in order to exert their influence on the world and shape it, he insists on the importance of embodying oneself in future scripts, as if one were rehearsing a play with an uncertain and uncontrollable storyline: "using scenarios is rehearsing the future" [SCH 91]. As Schwartz concedes, "there is an almost irresistible temptation to choose one scenario over the other: to say, in effect, *This is the future which we believe will take place. The other futures are interesting. But they're irrelevant. We're going to follow this scenario*" [SCH 91]. Yet the very purpose of scenario planning is to suspend our disbelief in a variety of futures that seem improbable, so that it ultimately becomes possible to prepare for the impossible [SCH 91]. As such, "the art of the long view" is a state of mind through which the unthinkable becomes apprehensible and its warning signs become recognizable. In short, it enables us to "plan for the future [and operate] in an uncertain world" [SCH 91], which is not the same as constructing the world we would want to operate in further along [SCH 91].

Following these and other principles that Schwartz calls "the art of the long view", the NAE's approach took place over 3 years, between 2001 and 2004, and mobilized a broad spectrum of people, backgrounds and disciplinary perspectives. Yet in the final report – *The Engineer of 2020* – the only trace remaining of these last theoretical principles is a formal evocation, which apparently contributes more to building credibility than to faithfully applying "the art of the long view". Among the principles discarded, those of *openness* and *non-investment* in scriptwriting are particularly revealing. They show how the narrative builds the confines of a transdisciplinarity that glorifies the engineering professions: in other words, it bounds a horizon of collaboration and knowledge whereby perspectives adhering to the foregoing principle of technological determinism are promoted onward, while others are excluded.

6.4.2. *The Engineer of 2020* or the "instrumentalization" of strategic forecasting theory

In *The Engineer of 2020* report, four prospective scenarios were developed. Each is described in a few lines in the introduction, while the full versions are relegated to an appendix, not to be referred to at any later point in the main text. Hence the challenge, under such circumstances, to track how much of the scriptwriting work followed "the art of the long view" it pretends to rely upon in order to get to the recommendations that make up the report. Notwithstanding its formal foundation in Schwartz's principles, the narrative emphasizes the intensity of technological change foreseeable between now and 2020, suggesting that no future is envisioned whereby technology would not be at the heart of everything. Therefore, at least one unavoidable scenario is missing from the picture: for what about the possibility – however unlikely – that technology is not the core nor the main determinant of solutions to current "grand challenges"? Whatever happens – are we to understand – the engineer's social role will have to be central in the future. And although it is formally stated that "the benefit of the scenario approach was that it eliminated the need to develop a consensus view of a single future and opened thinking to include multiple possibilities" [NAT 04], the main text nonetheless constitutes a single future, and even closed one: i.e. it constitutes a future that is ultimately built by and for the glory of engineers and engineering. As the discourse concedes, "the final two chapters express a bold optimism for the engineering profession if it is willing to confront the possibilities for the future and prepare for it" [NAT 14]. Contrary to Schwartz's principles [SCH 91], is it not indeed the closing on a scenario that is effected through them, as well as a broad recommendation for its investment by the American engineering profession?

In order to see how this *closure* and *investment* lead to a selective concept of transdisciplinarity – in the sense that it induces new boundaries more than it transcends old ones – we focus henceforth on the last two chapters, which describe a statement of aspirations for engineering in 2020, and a list of attributes that will enable the graduate engineer to live up to reality *as it will be* [NAT 04]. At first glance, the prevailing rhetoric is one that expands the engineering mind and improves upon how one becomes an engineer: in order to *rise to the occasion,* the engineer will have to aspire to a new vision and new skills[8]. Through new models such as "green engineering"[9], but above all

8 It is appropriate that engineers are educated to understand and appreciate history, philosophy, culture, and the arts, along with the creative elements of all of these disciplines. The balanced inclusion of these important aspects in an engineering education leads to men and women who can bridge the "two cultures" cited by the author C.P. Snow (1998). In our increasingly technological society, this is more important now than in the 1950s when Snow identified the issue. [...] Our aspiration is to shape the engineering curriculum for 2020 so as to be responsive to the disparate learning styles of different student populations and attractive for all those seeking a full and well-rounded education that prepares a person for a creative and productive life and positions of leadership [NAE 04].

9 Green engineering is the design, commercialization, and use of processes and products that are feasible and economical while minimizing the generation of pollution at the source and the risk to human health and the environment (National Science Foundation, 2003). [...] The

through "engineering without boundaries", we will have to avoid the pitfalls of specialized knowledge, which is acquired not only through a loss of scope, but also through losing sight of phenomena occurring at the boundaries between disciplines, precisely where opportunities for creativity and inventiveness emerge. Insofar as engineering is to become "without boundaries" [NAT 04], institutions will have to aim for their boundaries to be increasingly open, inclusive and generous. In order to meet the "challenge to an engineering future where interdisciplinarity will likely be critical to the solution of complex problems" [NAT 04], the engineers of the present must acquire the means "to rapidly embrace the potentialities offered by creativity, invention, and cross-disciplinary fertilization to create and accommodate new fields of endeavor, including those that require openness to interdisciplinary efforts with nonengineering disciplines [...]" [NAT 04].

In this way, the model of "engineering without boundaries" and the inter- or transdisciplinary opening that it implies refer to the foregoing "convergence" social imaginary: it aims at "integrat[ing] knowledge, tools, and ways of thinking from life and health sciences, physical, mathematical, and computational sciences, engineering disciplines, *and beyond*" [NAS 14, emphasis added], with a view to stimulating *disruptive* technological innovation, and thus offering a radically new paradigm for the resolution of today's "grand challenge" [DUP 04]. A plural epistemological otherness is called upon and included through this "beyond", which is guessed to correspond to the fields of social sciences and even art. But the question arises as to whether such "integration" at the boundaries [NAT 04] – also referred to as "disciplinary inter-pollination" by the NAE [NAT 04] – necessarily ends up evacuating a most crucial potentiality of inter- and trans-disciplinarity: i.e. that of revealing through art the fundamentally paradoxical character of digital technologies. In the words of Armand Mattelart, this would be saying that through such an "integration" at boundaries, "the existing naturalizes as an impassable horizon" [MAT 2009, authors' translation].

Although the seemingly generous internal and external boundaries of engineering are becoming blurred and increasingly permeable to projects in other fields [WIL 02, DOW 05], this is somehow happening in spite of

principles point to systems-based strategies and holistic approaches that embed social and cultural objectives into the traditional engineering focus on technical and economic viability [NAE 04].

engineering itself, under the "interfacing effect" of a "common digital language" instituted by computer engineering [WIL 02]. Williams [WIL 02] calls such a phenomenon "boundary dissolution", prior to illustrating its typical dynamic by the emergence of a Biological Engineering division at MIT and elsewhere, in engineering schools and faculties throughout the world:

> Nowhere is [this phenomenon] more evident than in the trading zone of biology and engineering. [...] A major factor in the success of this trading zone –and more generally a major factor in the rise of technoscientific mixes – is the role of information technology in providing a common, readily transferable language. […] The interfacing effect of a "common digital language" is seen throughout MIT, not just in biology-related areas. All engineering departments are becoming, in some form or other, to greater of lesser extent, departments of applied information technology. […] Almost all engineering departments are trying to recruit what are charmingly called 'computer science types' – adepts in the ligua franca of technoscience. Such types bring generalizing power to the department, but also tend to undermine departmental identity. It is hard to find someone who is expert in widely applicable computer-based systems and who is also interested in a particular industry or site associated with an engineering discipline. […] In the form of a common digital language, technology dissolves the familiar boundaries of engineering. It also lifts engineering, once the most down-to-earth of professions, from its familiar ground of materiality, endowing it with a new and ghostly lightness of being. Fewer and fewer faculty members in engineering actually make things or build things. More and more work with symbols and models. In a sense they are still working with machines, of course, but the meanings of both engineering and machinery are redefined when machines process information rather than matter. [WIL 02]

Yet the phenomenon stretches far beyond biological engineering [BEC 08]. In short, and more broadly still, the process of disciplinary integration at issue here is a process of ideological expansion that extends far beyond engineering, while at the same time being crucially dependent on it:

through this process, even choices that present themselves as so utterly reliant on politics devices (as opposed to "technical" ones) tend to surreptitiously turn into "problems to be solved". As a case in point, let us only think of the "subpolitics" [BEC 08] sustaining today's "smart city" paradigm: that is, a rapidly spreading paradigm whereby all cities already are "smart" or very rapidly becoming so by "the actions of millions of anonymous [smart phone] stakeholders who are experimenting with a new relationship to the urban environment in which the human and the non-human are becoming more entangled by the day" [PIC 15]. So indeed, "while 'technology' increases its rhetorical reach, that of 'engineering' shrinks" [WIL 02]; but above all, the rhetorical reach of the "problem-form" silently transcend the boundaries of engineering so as to infuse the Politics with the idea of a pure potentiality that is always extant, and to be collectively discovered. All therefore seems to occur as if *The Engineer of 2020* acted as an invisible symbolic mediator at the intersection between the two types of technological inevitability: in other words, after the "Metamorphosis of the bourgeois" [ELL 98] that occurred in the 20th Century, we would now be witnessing a metamorphosis of the engineer. As if *The Engineer of 2020* stood in the interstice that is both separating and connecting modern societies to technology, the institutional discourse underlying it (or its analogues) is itself imbued with the paradox and disseminates it: even though it is ontologically impossible for *The Engineer of 2020* to conceive of "technology" as autonomous, "technology" does become autonomous through its own narrative, in spite of all attempts at avoiding it; and even though it is ontologically impossible for the Politics to resolve in the manner of a technical problem, the latter nevertheless comes to encompass the former.

The rhetoric of technology and corresponding ideology of history would therefore seem to be expanding, i.e. technology no longer would appear as an element of history, as much as its very foundation [WIL 02]. Consequently, institutional engineering would be undergoing "expansive disintegration"; in the process of disappearing as a coherent and independent profession, it no longer would be able to define itself through well-understood relationships with industry, other social organizations and the material world. As such, "engineering is [indeed] ending", but "only in the sense that nature is ending: as a distinct and separate realm" [WIL 02].

Hence the ontological vertigo palpable in *The Engineer of 2020*, and beyond (to varying degrees depending on national contexts), within contemporary engineering professions. Yet such "engineering without boundaries" that embraces so broadly that it disintegrates becomes the Politics itself: thus subservient, engineering cannot conceive of itself as being embraced by "problem-form", nor of the latter as having a limit or a paradoxical foundation. From the point of view adopted in this chapter, this leads (or at least contributes) to the dominance of a one-way, or truncated interdisciplinarity where art – along with the perspectives in social sciences that mobilize it as a source of knowledge – can indeed contribute to problem-forming, but these problems nevertheless remain framed by others (namely engineers). It is in this sense that, while actually being invited more and more to processes of problematization, art is not granted as yet a full right of speech within them, with respect to the paradoxical experience it unlocks. Interdisciplinarity is put to work in the conceptual and ideological apparatus historically carried by the engineer – that is to say, in the apparatus of "problem-forming" and the pure potentiality that it embodies a priori – but as such it is confined to a utilitarian role that legitimizes it solely as an ability to respond to the contemporary injunction to produce solutions, while truncating it from its essential contribution. Killed or discredited even before they can reveal their power to re-symbolize the relationship between society and technology, the alternative voices and perspectives of interdisciplinarity such as art are no longer (or not yet) established as "lines of flight" (or *lignes de fuite*) [DEL 89].

6.5. Science fiction as emancipation from the "problem-form"

Ultimately, it is the relationship we have and want to have with "technology" that remains unquestioned by the "problem-form". The hypothesis of the inevitability of technological change in fact consists of removing technological development from the processes of politics orientation in order to give free rein to the discretionary choices of the actors who have control over it. Reflection is guided and determined by this presupposition of inevitability i.e.; but it is this same presupposition that invites, calls for and incites, and by the same token selects, proscribes, dismisses and sorts out the various proposals. So the space for reflection is conditioned and configured according to the prior adherence to the idea of a future that always is up for grabs, to be made and shaped collectively, but that must necessarily be technological. So conceived as inevitable, this technological change acts as a mechanism to control access to debates on major issues.

Hence the urgent need to reinvest social imaginaries beyond the "problem-form", so as to reopen possibilities obscured by discourses of technological revolutions. From this point of view, science fiction offers a potential as yet untapped by scientists and engineers [GEN 17]. Yet in order to let it be perceived, science fiction must itself be situated in the broader current of utopias that always have allowed the counterfactual question of "what if" to be expressed, and to expand the body of work from which scientists and engineers draw their inspiration.

As Musso recounts, utopia has changed in nature and meaning throughout history. For a long time, it has sent us back to a bygone golden age populated by gods [MUS 10]. With the discovery of America, it was the change of scenery that allowed More, Campanella or Bacon to stage ideal cities embodying scientific and politics utopias. In the 18th Century, utopia served as a vehicle for social criticism, but also embodied hopes for progress. As the Industrial Revolution began, several economic, social and politics utopias came about under the work of Jean-Baptiste Say, Charles Fourier, Robert Owen, Etienne Cabet and Saint-Simon [MUS 10]. But when the revolutions of the 1830s unfolded in France, these utopias were gradually replaced by a new generation of works: the technoscientific utopias.

The reason was that, in order to conceive of change and effect it, one inevitably had to decide between two politics-theoretical conceptions: change would either result from new social relations, or it would be induced by the development of technical networks: "to give primacy to politics or technology, to conflictuality or 'universal association', and finally to 'communism' or 'communication'" [MUS 10, authors' translation]. The *inflection* of utopias thus reflects a technocratic appropriation of power that evacuates social and socialist utopias in such a way that social change is reduced to technical achievements [MUS 10]. Politics conflict is obviated in favor of a celebration of technical progress and the industrial revolution, just as the railway system and telegraphic network are being structured [MUS 10].

The socialist way of change (propelled by politics utopia) has thus given way to the technical utopia of large networks, which have established themselves as tools of democracy and equality: "Improving communication

[...] is to generate equality and democracy. Sophisticated means of transport have the effect of reducing distances not only from one point to another, but also from one class to another" [CHE 38 cited by MUS 10, authors' translation]. Technical innovation thus makes it possible to avoid transformation via politics channels in a "modern myth [that] has imposed itself and will be reactivated at the time of each major network innovation, heralding a new social and economic revolution, from electricity to the Internet" [MUS 10, authors' translation]. By positing technology as a determinant of social change, the utopian problem drifted from the politics to the technical field [MUS 10, p. 8]. The Politics is thus overshadowed by a technocratic and financial power wielded by "engineers" whose faces are constantly multiplying:

> Politics of the future will have as its object the administration of the material interests of society; the general men of industry, the bankers and the engineers will then be politicians in at least the same capacity as the thinkers, the regulators. Our efforts must from now on be directed towards revealing the politics character within them, and that they do not feel. [Article in *Le Globe*, December 16, 1831, quoted by MUS 10, authors' translation]

As an "engineers' paradise", the new technological utopia – both normative and utilitarian an ideology – is operated through the "problem-form". It is embodied in the exploration of new worlds, in the development of new technologies, but also in caricatured politics architectures. According to Musso, they are less intended to criticize than to fascinate or frighten [MUS 10]. So much so that they have the effect of pitting one technophile and technophobic against the other. But more importantly:

> Techno-utopia has [...] become an ideology of technological fatality, this new great contemporary *fatum*. Technology has become the end and the means to think and achieve the social transformation and even revolutions of our time. It can be qualified as an ideological utopia: its movement is only the appearance of disruptions because it does not make use of utopias of social transformation.

> This is what fetishization is all about. There is no need to think of new social utopias, technological utopia would take charge of bringing about change [...].
>
> Criticism of this powerful ideological techno-utopia presupposes the triple rejection of fetishism, determinism and technicist progressivism. It would reject the reification of the technique that affirms its social exteriority before imposing it in return as the fatal causality of social and environmental upheavals. If a free population is a population that can still imagine something other than what it is, the challenge is to reinvent social, cultural and politics utopia against contemporary ideological techno-utopia. [MUS 10, authors' translation]

While it is true that technological utopias are proliferating [MUS 10], many science fiction works are on the fringes of this techno-utopia. They offer essentially politics reflections, which can feed a deeper or renewed understanding of social dynamics. This alternative interpretation is all the more effective in transforming social representations as it passes through the affects and immersive narration that the novel allows.

> The dramatic process is in this respect superior to the conceptual statement, since it alone gives rise to the feelings through which moral judgment is manifested. However, the role of emotions goes far beyond the moral framework since, as neuroscientist Antonio Damasio explains, affects are interwoven with cognitive processes so that the emotions that fiction evokes are part of a more general knowledge. By soliciting the unconscious and affective dimensions of judgment, the thought experiment made possible by fiction thus goes far beyond a purely intellectual exercise and nourishes a knowledge that is more integrated because it is tied up to emotions. [GEN 16, authors' translation]

Gender and feminist science fiction is a good example. By inventing single-sex worlds (such as Charlotte Perkins Gilman's *Herland* in 1915 or more recently Ursula Le Guin's *The Left Hand of Darkness* in 1969) or multiple-sex worlds, this literature has illustrated possibilities that have both challenged sexist relations and inspired an egalitarian perspective on gender[10]. According to Evans, despite the patriarchy that has long dominated gender (Vonarburg in Geekosphe), few mediums have been as effective in articulating the aspirations of feminism by portraying worlds free of sexism or utopias that transcend gender, speculations that can shake our certainties and preconceptions [EVA 14].

But it is the dimensions of discovery, adventure and conquest that are most often highlighted when science fiction is mentioned, which reinforces not only a technicist perspective, but also a male perspective of science fiction [GEE 18]. In the process, the rational perspective, precision and technology of the works are emphasized at the expense of the moral, politics and symbolic exploration that narrative immersion in an alternative world allows. While the novels of Joanna Russ, Marge Piercy, Ursula K. Le Guin and Octavia Butler accompanied the feminism of the late 21st Century, even a work like Star Trek questioned race relations, imperialism and the Cold War [EVA 14].

Some commentators argue that science fiction works by women often offer a very reflective dimension of society [GEE 18]. Feminist works have staged dystopian universes such as *Swastika Night* by Katharine Burdekin or Margaret Atwood's *The Handmaid's Tale*. In recent years, several authors have contributed to the revival of essentially social and politics dystopias: Suzanne Collins' *Hunger Games* or Veronica Roth's *Divergent*, SJ Kincaid's *Diabolic*, Cecelia Ahern's *Destiny*, or Naomi Alderman's *The Power*.

> Some of the stories give pride of place to the anthropological and sociological study made possible by the novel of anticipation, the beauty of the 'and if it were different' that Elisabeth Vonarburg loves so much. Mary Doria Russel will not hesitate in *The Sparrow* to evoke the culture of the extraterrestrials encountered by her heroes. Cultural aspects, traditions and language are accurately dissected, the author being herself an anthropologist. Much of Ursula Le Guin's work is based on this same attention to detail, this same observation of society, which is particularly visible in *The*

10 Evans cites Joanna Russ' *The Female Man* 1970, Marge Piercy's *Woman on the Edge of Time*, Jayge Carr's *Leviathan's Deep*, Sally Gearhart's *The Wanderground*, Suzy McKee's *Charnas* and *The Left Hand of Darkness* by Ursula K. Le Guin.

> *Left Hand of Darkness*. [The Geekosophist, 2018, authors' translation and use of emphasis].

But male authors also explore human nature and social relationships, as in Ted Chiang's *The Story of Your Life* (1998) adapted for the cinema by Denis Villeneuve in *Arrival* (2016) or Scott Westerfeld's *Uglies* series about the dictates of beauty. In short, science fiction is full of works that can emancipate the scientist and engineer from a technicist perspective of society and future. There is therefore every interest in better highlighting the sociopolitical corpus of science fiction, as well as insisting on a politics-social interpretation of works that are nevertheless part of a techno-utopic trend. This is all the more true since, far from the optimistic techno-utopias in which it was first embodied, science fiction today delves deeper into the social effects of technology by portraying societies that are still experiencing redefined social relationships. On the other hand, science fiction can certainly be mobilized in the art of the long view to which Schwartz invites us.

6.6. Conclusion

While the "problem form" historically and archetypically ties to modern engineering, it now prevails on a much wider scale, and well beyond its initial archetype. The corollary hypothesis of the inevitability of technological change has the effect of removing technological development from the processes of politics orientation so as to give free rein to the choices of the actors (experts) who control it; but it also feeds into processes which, under the guise of a superficial epistemological pluralism, reject the heuristic premises of disciplines that would allow us to think differently about social issues and the grand challenges facing humanity.

Even when art and the social sciences are mobilized in a process of interdisplinarity, the space for reflection is conditioned and configured according to the "problem-form": i.e. the precondition of adhering to the idea of a future conceived as a "problem to be solved" discredits their most radical, and potentially also most fruitful, contributions. The technological solution operates as a mechanism to control access to debates on major collective issues, to which both art and social sciences must submit to if they wish to have a voice. Thus, the opening that is actually taking place through the most innovative modalities fails to (re)symbolize reality. However, science fiction could act as a powerful heuristic for renewing our views and unlocking new potentials as yet prohibited by the problem-form and the hypothesis of technological inevitability [DUP 04].

By conceiving of technology as a main determinant of social change, the utopian problem had closed in on a binary oscillatory movement between the social order and its binary opposite: disorder. Yet following this initial movement from the politics to the technical [MUS 2010], today's utopia is taking over sociopolitical issues that cannot be reduced to problem-solving or technological solutions; it crucially depicts a new sort of technocratic and financial power, whereby engineers often find themselves trapped in paradoxical positions.

6.7. References

[BEC 08] BECK U., *La société du risque: Sur la voie d'une autre modernité*, Éditions Flammarion, Paris, 2008.

[BRA 05] BRADFIELD R.W.G., BURT G., CAIRNS G. *et al.*, "The origins and evolution of scenario techniques in long range business planning", *Futures,* vol. 37, pp. 795–812, 2005.

[CAB 11] CABANTOUS L., GOND J.-P., "Rational decision making as performative praxis: explaining rationality's *Éternel Retour*", *Organization Science,* vol. 22, no. 3, pp. 573–586, 2011.

[CAR 10] CAREW A. L., WICKSON F., "The TD wheel: a heuristic to shape, support and evaluate transdisciplinary research", *Futures,* vol. 42, pp. 1146–1155, 2010.

[DEL 89] DELEUZE G., "Qu'est-ce qu'un dispositif", in *Michel Foucault Philosophe: Rencontres internationale 1988,* Association pour le Centre Michel Foucault (eds.), Le Seuil/Des travaux, Paris, pp. 185–195, 1989.

[DOB 15] DOBBS R., MANYIKA J., WOETZEL J., The Internet of Things: Mapping the Value Beyond the Hype, McKinsey Global Institute (MGI), San Francisco, 2015.

[DOW 05] DOWNEY G., "Are engineers losing control of technology? From 'problem solving' to 'problem definition and solution' in engineering education", *Chemical Engineering Research and Design,* vol. 83, no. A6, pp. 583–595, 2005.

[DUP 04] DUPUY J.-P., "Quand les technologies convergeront", *Revue du MAUSS,* vol. 23, pp. 408–417, 2004.

[GEN 16] GENDRON C., AUDET R., "L'environnement au prisme de la science-fiction. De la dystopie à la reconstruction du rapport Homme-Nature dans les sociétés postécologiques", *Futuribles*, July-August 2016.

[GEN 17] GENDRON C., IVANAJ S., GIRARD B. *et al.*, "Science-fiction literature as inspiration for social theorizing within sustainability research", *Journal of Cleaner Production*, vol. 164, pp. 1553–1562, 2017.

[GRA 03] GRAZ J.-C., "How powerful are transnational elite clubs? The social myth of the world economic forum", *New Politics Economy*, vol. 8, pp. 321–340, 2003.

[IAN 14] IANSITI M., LAKHANI K.R., "Digital ubiquity: how connections, sensors, and data are revolutionizing business(digest summary)", *Harvard Business Review*, vol. 92, no. 11, pp. 91–99, 2014.

[JAR 05] JARZABKOWSKI P., *Strategy as Practice: An Activity Based Approach*, Sage, New York, 2005.

[KER 13] KERELUIK K., MISHRA P., FAHNOE C. *et al.*, "What knowledge is of most worth", *Journal of Digital Learning in Teacher Education*, vol. 29, pp. 127–140, 2013.

[KLE 14] KLEIN J.T., "Discourses of transdisciplinarity: looking back to the future", *Futures*, vol. 63, pp. 68–74, 2014.

[KOH 18] KOHN RÅDBERG K., LUNDQVIST U., MALMQVIST J. *et al.*, "From CDIO to challenge-based learning experiences—expanding student learning as well as societal impact?", *European Journal of Engineering Education*, vol. 45, no 1, pp. 22–37, 2018.

[LAY 86] LAYTON E.T., *The Revolt of the Engineers: Social Responsibility and the American Engineering Profession*, Johns Hopkins University Press, Baltimore, 1986.

[MAT 09] MATTELART A., *Histoire de la société de l'information*, La Découverte, Paris, 2009.

[MOT 16] MOTE JR. C.D., DOWLING D.A., ZHOU J., The power of an idea: the international impacts of the grand challenges for engineering, *Engineering*, vol. 2, no. 1, pp. 4–7, 2016.

[MUS 10] MUSSO P., *Saint-Simon, l'industrialisme contre l'Etat*, Nouvelles éditions de l'Aube, La Tour d'Aigues, France, 2010.

[NAT 04] NATIONAL ACADEMY OF ENGINEERING, *The Engineer of 2020: Visions of Engineering in the New Century*, National Academies Press, Washington, DC, 2004.

[NAT 08] NATIONAL ACADEMY OF ENGINEERING, *The Grand Challenges for Engineering*, National Academies Press, Washington, DC, 2008.

[NAT 14] NATIONAL ACADEMY OF SCIENCE, *Convergence: Facilitating Transdisciplinary Integration of Life Sciences, Physical Sciences, Engineering, and Beyond*, National Academies Press, Washington, DC, 2014.

[PIC 15] PICON A., *Smart Cities: A Spatialised Intelligence*, John Wiley & Sons, Chichester, 2015.

[SCH 16] SCHWAB K., The Fourth Industrial Revolution, World Economic Forum, Geneva, 2016.

[SCH 91] SCHWARTZ P., *The Art of the Long View*, Currency Doubleday, New York, 1991.

[VAR 10] VARUM C.A., MELO C., "Directions in scenario planning literature – a review of the past decades", *Futures*, vol. 42, no. 4, pp. 355–369, 2010.

[VER 93] VERIN H., *La gloire des ingénieurs: l'intelligence technique du XVIe au XVIIIe siècle*, Éditions Albin Michel, Paris, 1993.

[WIL 02] WILLIAMS R., *Retooling: A Historian Confronts Technological Change*, MIT Press, Cambridge, 2002.

[WIS 12] WISNIOSKI M., *Engineers for Change: Competing Visions of Technology in 1960s America*, Massachusetts Institute of Technology Press, Cambridge, 2012.

7

Design Fiction, Technotypes and Innovation

7.1. Introduction

The use of designers to create prototypes is a widespread practice in research and development centers. Innovators are increasingly interested in science fiction, which offers utopian technologies that are very useful in representing the goals to be achieved by engineers. Designers are involved at different levels of the innovation process [MIN 16]. They are also called upon, along with engineers and scientists, to advise directors of science fiction films that are supposed to be made in the future. Interactions between science fiction and R&D are increasingly frequent, giving rise to emulation and a virtuous cycle [MIC 10a]. Most of the great science fiction films offer visions of the future with many technological details. The gadget dimension of science fiction contributes to its success. A film such as *Back to the Future II* (1989), which takes place in the future (in 2015), anticipated the appearance of many inventions that became commercial goods several years after their theatrical release. In the other direction, artists and designers can help companies imagine tomorrow's world in order to better prepare the insertion of their future products in markets often saturated with novelties. Building a vision of the future is necessary for business strategy. For this, decision-makers have become accustomed to soliciting artists, especially science fiction writers. They contribute to building visions of the future that will ensure the success of future strategies. Also in ancient times, the leaders of society used oracles to validate their ambitions and plans for the city

---

Chapter written by Thomas MICHAUD.

[CAZ 86]. Politics and divination have long worked together. Today's decision-makers are increasingly turning to science fiction, which has a reputation for its forward-looking descriptions [MIC 10b].

The vision of the future of the technoscientific era is essentially science fiction. Over the past century or so, many authors have contributed to feeding society with futuristic fictions whose influence on the collective imagination is evident [BRA 08]. The vision of the future is therefore important for a State, a company, or for any institution seeking to convince the markets of its merits. Companies must therefore appropriate a science fiction imaginary world that certainly corresponds to the times, but whose aspects, such as the violence of certain scenes, can put off a certain audience. The need to propose a vision that is both consensual and original leads the most innovative companies to solicit the contribution of science fiction authors, or coaches specialized in this subject, to stimulate the imagination of their collaborators, during new kinds of creativity sessions, often generating amazing results. Brainstorming is outdated, and design thinking has transformed into design fiction, i.e. the stimulation of the individual or collective imagination, with the aim of generating prototypes, in the form of objects or fictions, which will then be used to think about strategy and R&D activities in a better way [DUN 13].

So science fiction is no longer just an issue for the cultural industry. Its technopolitical discourse is of paramount importance, as it tends to structure the representations of the future of some political leaders, to the point of marginalizing the visions of political decision-makers, who are sometimes less influential than some films. The public assimilates the major technological and political trends of humanity and forges a vision of the future of science fiction, which is supposed to validate the research programs of the most powerful actors of the technoscientific era, i.e. technology companies, of which GAFAM are the most emblematic examples. The latter show their interest in science fiction by publishing anthologies of institutional science fiction, or by organizing science fiction competitions. Design fiction is even making its appearance in forecasting firms, or in the strategic management and R&D departments of large companies. Everyone wants to develop visions of the future of science fiction, aware of the impact of this performative imagination on employee motivation, and then on the markets to be conquered [MIC 17].

Design fiction intervenes as a means of filtering the imaginary world of science fiction in order to conform it to the ethics and values of the company or institution. The blockbuster scenarios are thus hijacked and inserted into systems corresponding to the moral codes of technocapitalism. The company mobilizes its employees to assimilate the science fiction imaginary world into stories that we call *institutional science fiction*.

The technoscientific era is characterized by the development of science that is more applied than fundamental, in the service of a technology at the center of productive interactions. Capitalism thus uses the imaginary world as a means of justifying its system, creating mobilizing futuristic visions [HOT 13]. It is not surprising that futuristic science fiction comes mainly from the United States [ALK 10]. This country is the most powerful and it communicates its vision for the future of humanity in this way. On the other hand, it is a new country with a very recent history. It compensates for its lack of history compared to Europeans or Chinese, for example, by proposing visions of the future, in which it projects itself and envisages its evolution [BAU 98]. American science fiction, since Hugo Gernsback, has proposed thousands of utopian technologies. But the United States is not the only country that appreciates science fiction and uses it to project itself into the future. The British, the French and the Europeans laid the foundations for this kind of work as early as the 19th Century [WES 98]. The Russians, then the Soviets, also produced futuristic works of fiction. The Japanese have experienced a period of anticipatory science fiction, particularly through manga, especially since the 1980s. Finally, China, increasingly powerful, also appreciates science fiction, without considering it as an imaginary world specific to American capitalism.

This chapter will first present pioneering approaches to creativity through science fiction, before developing some features of design fiction. Finally, the theory of technotypes will be discussed. It is a question of considering how the imaginary world, and mainly the utopian technologies of science fiction appear, drawing inspiration from a collective unconscious within which technological archetypes, or technotypes, function. More generally, Altshuller, science fiction author and inventor of the theory of the resolution of invention-related tasks (TRIZ) method of creativity, suggests using these works of fiction to stimulate the imagination of scientists and engineers.

7.2. Altshuller, from science fiction to the TRIZ method

> Reading science fiction undoubtedly helps to develop the creative imagination; however, it cannot replace systematic training. Imagination must be systematically developed through special exercises[1].

Genrich Altshuller devoted part of his research to the impact of science fiction and fantasy on the creativity of engineers and scientists. His TRIZ method is used by the most innovative companies. In the book *The Innovation Algorithm* [ALT 07], he wondered why fantasy was not used more:

> It has become a textbook maxim that fantasy plays a large role in any creative activity, and in technical science as well. But there is a surprising paradox here. The recognition of fantasy's importance has not been accompanied by a systematic effort, focused towards its development.
>
> So far, the only widespread, and practically effective means for developing fantasy was the reading of science fiction literature (SF). Incidentally, a clear correlation is seen here: scientists and engineers are more attracted to SF than other readers. Several years ago, the Committee of Technical-Scientific Literature of the Azerbaijan Writers Union conducted a survey (…) Fifty-two percent of the engineers and physicists surveyed mentioned that they value SF first of all for its new technical-scientific ideas. Really, in this regard, SF can give the thinking engineer quite a lot: a project that can be developed, or even a ready solution that can be transferred into engineering language[2].

The researcher, author and science fiction enthusiast quickly became aware of the importance of the imaginary world on the creativity of scientists. Although science fiction developed largely in Europe and the United States, it is important to note that this genre also found favorable

1 Altshuller Genrich, *The Innovation Algorithme, TRIZ, Systematic Innovation and Technical Creativity*, Technical Innovation Center, Inc., Worcester, MA, 2007, p. 232.

2 Altshuller Genrich, *The Innovation Algorithm TRIZ, Systematic Innovation and Technical Creativity*, Technical Innovation Center, Inc., Worcester, MA, 2007, p. 229.

ground in the USSR. Altshuller was a connoisseur of the history of technology and patents. He contributed to the study of the technical imagination and analyzed hundreds of utopian technologies. He was aware of the impact of science fiction on the scientists and engineers of his time. Thus, he established a cultural link with the United States, a rival country considered the promised land for technoscientific extrapolations and thought experiments. Thus, the dominant countries developed this imagination for several decades, to the point where gender appeared as a godsend for creativity researchers. Propelling a researcher into an imaginary world with a science fiction connotation already appeared in the 1960s and 1970s as a necessary step in the development of his faculties and productivity. Science fiction was no longer just a distraction, but also a serious genre, capable of anticipating the most decisive innovations:

> Very often, the ideas of fiction writers are directly used during the early development stage of a new field of science and technology. At some period (although, for a very short time) fiction becomes one of the main sources for an emerging new area of knowledge[3].

New sociotechnical paradigms are prepared in an imaginary world that evolves as the inventions it helped to represent become innovations. Altshuller confirms that science fiction and the imaginary world develop upstream of the innovation process, diffusing utopian technologies into the imaginations of engineers who are often specialized in sectors looking for futuristic perspectives. Scientific communities are sometimes small during research and development work, before growing when fundamental discoveries are made and seek to become commercialized innovations. The existence of imaginary narratives dealing with their field of investigation can be crucial in order to unite the research community and then to publicize to the general public the potential applications of the discoveries made. The author reminds us, however, that we should not succumb to a blissful faith in the creative potential of science fiction. Many stories are indeed of poor quality, which contributed to the denigration of the genre by the intelligentsia for several decades. However, it is interesting to note the evolution of this old "bad genre", which ended up converting the elites at the same time as the technosciences became more and more legitimate and

3 Altshuller Genrich, The Innovation Algorithm TRIZ, Systematic Innovation and Technical Creativity, Technical Innovation Center, Inc., Worcester, MA, 2007, p. 230.

then valued by the new masters of Silicon Valley. Scientists became increasingly powerful after the Second World War, particularly under the impact of the cybernetic paradigm [LAF 04]. However, science fiction did not cease to spread among scientists, who drew inspiration from it at all times:

> Of course, science fiction does not always contains correct and mature ideas. Often, they are most doubtful from a scientific – technological point of view. Or, they are completely symbolic ideas offered to readers. Moreover, often fictional ideas are completely wrong. In spite of that, because of their singularities and brilliance, they attract the attention of researchers, and force intensive research that sometimes leads to important discoveries or inventions[4].

Altshuller was convinced of the innovative potential of science fiction to go beyond the current and contingent knowledge of science. It would even be a form of madness, in the positive sense of the word:

> SF helps overcome psychological barriers on the road to 'crazy' ideas without which science cannot continue its development. This is an admirable, and so far, little acknowledged function of SF that becomes a component of the professional training of scientists[5].

Without this subversive imagination, science could be condemned to stagnation and the accumulation of knowledge without paradigmatic disruptions. Innovation processes mobilize a multitude of factors, and imagination could be considered a marginal element, or even a danger to the rationality of inventors. That is not the case. Science fiction has accompanied industrial revolutions since the 19th Century and contributes to questioning, criticizing and stimulating scientific discoveries [BAR 07]. It is a tiny and yet necessary factor for innovation. Without it, major changes would be difficult to conceive of. Altshuller's approach confirms the hypothesis that the imagination must be stimulated during the training of

---

4 Altshuller Genrich, The Innovation Algorithm TRIZ, Systematic Innovation and Technical Creativity, Technical Innovation Center, INC. Worcester, MA, 2007, p. 230.

5 Altshuller Genrich, The Innovation Algorithm TRIZ, Systematic Innovation and Technical Creativity, Technical Innovation Center, INC. Worcester, MA, 2007, p. 230.

engineers and scientists, but also in companies and R&D centers, when looking for disruptive innovations or solutions to certain problems.

Altshuller studied science fiction carefully in order to reach these conclusions, which are integrated into a more global reflection on the development of creativity and technology in history:

> It is wrong to say that SF is an irreplaceable creative tool for science and technology. However, it is, without doubt, one of the most important tools. The recording and careful analysis of SF ideas is long overdue. (...) In 1964, I started to create a Registry of Contemporary Science Fiction Ideas. Today almost all interesting ideas are registered in the list. They are separated into 12 Classes, 75 sub-classes, 406 groups, and 2,360 sub-groups. This analysis answers the question: 'When has a fictional idea become successful, and when has it not?' Moreover, some of the patterns in the generation of fictional ideas become clearer.

As early as the 1960s, the author began to classify the science fiction imaginary world. Since his death, such an approach has been replicated by the Website www.technovelgy.com. More than 2,000 science fiction inventions are compared with the history of technology in order to establish reflections on the interactions between science and science fiction. The science fiction author and theorist Brian Stableford has compiled an encyclopedia on the subject [STA 06]. Does it make sense to compare the history of utopian technologies with the history of techniques, and is it valid from a methodological point of view? It would be an exaggeration to consider that all technical inventions are preconceived by science fiction in the first place. Yet studies by Altshuller and Technovelgy show particularly disturbing regularities. According to these approaches, the imaginary would have a central place in the processes of innovation that need to be theorized and analyzed. Where does this technical imagination come from? How does it spread in society? What is its impact on engineers and scientists? More generally, is there a collective imagination, or even a technological unconsciousness in which there are the technotypes that humanity has been translating into innovations since the beginning of history in the form of archaeotechnics, and then inventions whose accumulation raises the question of the purposes of technology?

In order to optimize innovation processes, the ideal would be to recognize the purpose of technology, conceived as a global phenomenon that should lead humanity towards unknown trajectories. Altshuller believes that technical systems are evolving in response to regularities. The laws behind the TRIZ were established following analysis of 1,500 of these systems [ALT 64]. Patents ensure the crystallization of knowledge and are organized in a substantially equivalent way in most dominant technical systems. Cavalucci [CAV 16] believes that engineers should use these laws in their prospective thinking and invention activity. He adds that "the role of the inventor would consist of being only the opportune craftsman who accompanies the passage of the object from one generation to another on a generic pre-written line"[6].

Altshuller introduced a reflection on the knowledge of the regularities, laws and purposes of technology, which he considered attainable through a scientific approach. The TRIZ method has proven its effectiveness many times. However, we will retain Altshuller's theories only those devoted to the use of science fiction to innovate. So how does the imagination influence the R&D process? If there are laws establishing the evolution of technical systems, what are the laws governing the technical imagination in contact with engineers and scientists?

We have studied in several publications [MIC 18] the functioning of the imaginary world of virtual reality, in particular through the analysis of about 50 works of fiction. While this innovation has its origins in technotypes, it is constantly evolving and accompanied by utopian technologies, such as William Gibson's cyberspace (*Neuromancer*, 1983), Neal Stephenson's metaverse (*Snow Crash,* 1992), the Wachowski brothers' matrix (*The Matrix,* 1999) and Cline's OASIS (*Ready Player One*, 2011) for the most famous, which ensure the transition from one stage of this technical system to another. Utopian technologies appear, stimulating the imagination of engineers, and then becoming outdated and eventually becoming commercialized realities [MIC 08]. It is then that new utopian technologies emerge, and boost R&D, seeking to realize the technical imagination. While many thinkers have sought to define the laws that underlie the mechanism of the historical dynamics of technology, we propose to investigate the

---

6 Cavalucci Denis, "Genrich Altshuller. Theoriser l'acte inventif pour mieux l'enseigner", in THIERRY B.-H., CAROLINE H., PATRICK C. (eds), *Les grands auteurs en management de l'innovation et de la créativité,* Editions EMS, 2016, Paris, p. 565.

regularities and functioning of the technical imagination mainly through the study of the utopian technologies of science fiction. Under these conditions, the mapping of the technical imagination should make it possible to propose ever more effective methods of creativity, and to guide teams or individuals in search of innovations toward the works of fiction most likely to be integrated into the R&D program of the institution that requests them. Two types of mapping are thus recommended. The first is interested in the popular, collective imagination, notably produced by the cultural system, made up of novels, films and comic books, some of which have a global impact. The second focuses on technical fictions produced within an institution as part of creative sessions such as brainstorming or design fiction. We will evaluate later on how this latter approach works and is likely to contribute to a dynamic spiritualization of the companies and organizations that request it.

7.3. John Arnold's approach

Altshuller evoked Stanford Professor John Arnold's approach, who used a method inspired by science fiction to stimulate the creativity of young researchers. He proposed to project himself into the environment of an imaginary planet, Arcturus IV, in order to project problems and establish original solutions. This star is different from the Earth in terms of climatic conditions and geology. For example, its atmosphere is made up of methane and the oceans of ammonia. Gravity is 10 times stronger than on Earth and the intelligent beings that inhabit it are birds. The goal is to break down psychological barriers to design, for example, cars, houses, and technologies adapted to this fictional environment. Arnold's goal was to stimulate the creativity of engineers in order to make them more competitive. At the beginning of this initiative, he faced some criticisms which were however quickly dispelled:

> Science fiction in the classroom? What! You're designing for non-humans on far distant planets? Aren't there enough unfilled human needs that you could design for and thereby better use your time? These are some of the typical questions that are asked when people first hear of the Arcturus Project used in the Product Design at MIT. After explaining the project and the course, however, questions usually change to exclamations such

> as, 'What an idea, I wish I could have taken a course like that!'[7].

It is interesting to note that this approach was being developed in one of the most symbolic places of American science and innovation, MIT (Massachussets Institute of Technology), at a time when cybernetics was enjoying a certain success in the academic world [WAR 82]. At the same time, science fiction was taking advantage of the craze for hard science to become increasingly popular, to the point that it constituted a cognitive derivative in the service of the most rationalist theories. Since the end of the Second World War, science and science fiction have continued to flourish and gain institutional power. Very early on, some American pioneers like John Arnold understood the value of the imaginary world in stimulating creativity [BIL 14]. The stakes were high, as it was a matter of shaping the minds of the future researchers and elites of the American nation. At the same time, some actors, particularly from the psychiatric field, denounced the moral perversity of science fiction stories, but failed to slow down the enthusiasm of the scientific community for a genre interested in science, to the point of ensuring its prodigious growth for several decades. Better still, to develop the imagination, the use of science fiction was seriously considered as early as the 1950s:

> It is possible to train student to think more creatively; one can develop his imagination. The most encouraging aspect of the experiment is that this is as equally true for many who originally thought that they had little talent for design as it is for those who had previously exhibited a high order of imagination. The students claim that they leave the course with a new perspective with which to face a broad variety of problems[8].

John Arnold was a pioneer in the philosophy of creative engineering. He noted that science fiction was particularly appreciated by this social group and deduced that it was possible to combine the useful with the pleasant by

---

7 Arnold John, *The Acturus IV Case Study*. The case is freely available on the Internet. At the end of the document is an article by John Arnold entitled "Space, Time and Education". The following page numbers are taken from this document circulated by John Arnold's son, p. 129

8 Arnold John, *The Acturus IV Case Study*. The case is freely available on the Internet. At the end of the document is an article by John Arnold entitled "Space, Time and Education". The following page numbers are taken from this document circulated by John Arnold's son, p. 129.

proposing a 3-week seminar during which young researchers could push back the limits of a sometimes exacerbated rationality at the expense of a creativity that was nonetheless useful for innovation. He presented this approach as a strategic issue for the future of the American nation. MIT, where John Arnold's classes took place, was also home to one of the world's largest science fiction libraries, with over 60,000 books collected through the MIT Science Fiction Society, established in 1951. Symbolically, this establishment has been combining science and science fiction since the 1950s, indicating that the two disciplines are working together on creative and imaginative technical progress:

> The aim, then, of the MIT Creative Engineering program is to provide an even increasing number of young men trained, not only in the basic concepts of science and engineering but also in the use of their creative imaginations, to help solve the ever increasing problems, both in complexity and in numbers, that continue to face the nation and the world. Design courses provide and ideal vehicle for this kind of training, but by no means should this training be restricted to this field[9].

John Arnold's original approach was met with great success in the 1950s and was supported by industrial sponsorships, beginning the process of legitimization of science fiction by American elites:

> And lastly, a great many of the students with imagination are already science fiction fans or else take to it very readily. The result is that the first case he works on is fun and not work; he learns while he enjoys himself. There may be some theory that education must be solemn and serious but the Creative Engineering Group at MIT do not subscribe to it. The results of the informal seminars and lab sessions indicate that it would be desirable to hold all classes in a similar fashion (…) The program in Creative Engineering is expanding rapidly. This is due in part to the encouraging interest shown by all industries aware of the work that is being done. A numbers of grants have been received

9 Arnold John, *The Acturus IV Case Study*. The case is freely available on the Internet. At the end of the document is an article by John Arnold entitled "Space, Time and Education". The following page numbers are taken from this document circulated by John Arnold's son, pp. 131–132.

> to be used for the preparation of new case material and other research and in one instance a large corporation went to considerable trouble and expense in the preparation of a very complete case history for one of the projects[10].

John Arnold paved the way for the increased use of science fiction in education and institutions. Many companies and administrations have since been inspired by his approach to set up programs to stimulate the creativity of their employees.

7.4. The emergence of design fiction

Design fiction is a practice that has been developed increasingly in organizations that use the science fiction imaginary world in order to innovate. The term was introduced by cyberpunk science fiction author Bruce Sterling and further defined more precisely by Julian Bleecker in a 2009 essay [BLE 09]. Bruce Sterling gave an interview for the *Slate* Website in March 2012 where he clarified the concept of design fiction: "It's the deliberate use of diegetic prototypes to suspend disbelief about change. That's the best definition we've come up with. The important word there is *diegetic*. It means you're thinking very seriously about potential objects and services and trying to get people to concentrate on those rather than entire worlds or political trends or geopolitical strategies. It's not a kind of fiction. It's a kind of design. It tells worlds rather than stories"[11].

Etienne Souriau defines diegesis as "everything that is supposed to happen, according to the fiction presented in the film; everything that this fiction would imply if it were assumed to be true"[12]. Design fiction makes it possible to represent futuristic objects, for example in videos. It is an addition to science fiction, experimental and more adapted to innovation processes imagining very specific futures, specific to particular economic

10 Arnold John, *The Acturus IV Case Study*. The case is freely available on the Internet. At the end of the document is an article by John Arnold entitled "Space, Time and Education". The following page numbers are taken from this document circulated by John Arnold's son, p. 141.

11 Bosch Torie, "Sci-Fi writer Bruce Sterling explains the intriguing new concept of design fiction", *Slate*, March 02, 2012, available at: https://slate.com/technology/2012/03/bruce-sterling-on-design-fictions.html.

12 Souriau Etienne, "La structure de l'univers filmique et le vocabulaire de la filmologie", *Revue internationale de filmologie*, no. 7–8, p. 240.

sectors. Design fiction allows a diversification of sociotechnical concepts through original works produced in an innovative framework. Designers are taking over from science fiction writers while borrowing their creativity in the technological field.

Bruce Sterling is an award-winning American popular science fiction author. He is considered one of the founders of the cyberpunk movement along with William Gibson [CAV 01]. This movement has contributed to the popularization of a futuristic technical imagination accompanying the emergence and diffusion of ICTs, biotechnologies and nanotechnologies [PAR 96]. Cybernetics is a recurring theme in these works of fiction, produced mainly during the 1980s and 1990s. Cyberspace is the flagship technology in that it prefigures a computer network described as a "consensual hallucination" by Gibson in the novel *Neuromancer*, heralding the creation of the Internet and its transformations toward achieving a global virtual world. This trend was considered avant-garde, particularly appreciated in the hacker and computer scientist communities that developed ICT in Silicon Valley and in R&D labs around the world [FLI 01]. Gradually, this imaginary world became popular and mainstream as films and novels attracted a growing audience. Like many alternative and subversive cultures, cyberpunk has become normalized and trivialized. Computer fans have developed the Internet and many innovations that have changed the world and the way we communicate. In the years 2000–2010, science fiction even became the medium for creative sessions. Design fiction is the ultimate manifestation of the trivialization and normalization of this imaginary world, assimilated by a large part of the population, not just a few enlightened participants. The cyberpunk imaginary world has reached a new stage, and it is symbolically interesting to note that it is Bruce Sterling, the founding father of the movement, who is at the initiative of this new creative method.

Since then, many researchers have written articles on the subject, leading many companies to develop expertise on science fiction and creativity sessions based on the stimulation of this imaginary world. Design fiction initially aims to provoke the emergence of new concepts through individual and collective fiction. Moreover, the development of prototypes is the ultimate goal of this approach, which streamlines organizational imagination to shape potentially interesting and innovative objects. Design can generate technological artifacts, but also short stories, videos, comic books, documentaries, catalogues or newspapers. The followers of this practice are always coming up with new approaches to project the imaginary world and crystallize it into imaginations useful for illustrating R&D projects or even strategic discourses [KIR 10].

With design fiction, the debate on the prophetic role of science fiction is revived [SCH 94]. Inventors use this technique to project themselves into the future, where they imagine potentially revolutionary scientific techniques or theories. Technological prototypes, or artifacts, feed the initial thinking of the projects. In industrial sectors, making a prototype is a very useful initial step to represent a product with a view to mass distribution. Marketing, R&D and communication departments, among others, can use these prototypes to represent the future of the company and optimize development and marketing strategies and policies from concrete elements [DUN 13].

By creating works of fiction, it is also possible to understand social and political phenomena that can influence business and the productive system. Design fiction does not aim to predict the future, even though its narratives often take place there. This perspective, however, has damaged the reputation of science fiction, which is accused of a divinatory ambition contrary to instrumental rationality. Pseudosciences very often assume this prophetic role of science fiction [GRE 94]. Many people believe in the anticipatory ability of this artistic genre, even if this futuristic dimension of their work is regularly contested and condemned by the authors [HOL 02]. Design fiction will thus stimulate the imagination of strategists. Creating a fiction of conflict between actors, for example between economic competitors or political adversaries, allows us to envisage the worst and ensure anticipation of possible attacks. Science fiction speculation is particularly useful in preparing for war or a business strategy. Staff and strategic committees therefore have a common interest in opting for this practice.

Thus, organizations develop design fictions that help structure their identities. They stimulate their imagination to create prototypes and storytelling that can have a structuring effect internally, but also externally, with regard to potential clients [SAL 07]. R&D needs such practices to structure corporate identity and strategic marketing is increasingly interested in the development of a science fiction and speculative rationality in the world [BUD 00].

Science fiction authors such as Neal Stephenson wonder why dystopias have been on the rise since the 2000s, fearing that they are no longer fulfilling their role as positive stimulators of industry and the economy. The writer was, however, employed by a large Silicon Valley company, Magic Leap, to become a futurist. While independent science fiction is also dystopian, it may be because positive narratives, made up of science fiction and utopian technologies, are developed within institutional structures. The most creative authors are employed to put their imagination at the service of economic interests. This approach could explain the dystopian trend of the years 2000–2010. It is likely that this orientation will be reversed as design fiction practices are externalized. After developing their works of fiction and extrapolating strategic scenarios for several years, organizations have the ambition to disseminate their identity to the general public and consumers. The concepts emerging from design fiction will then feed the markets and independent science fiction could take advantage of this to renew itself and offer utopian fictions again. Similarly, companies may fund artists to create science fiction anthologies that are distributed free of charge to potential clients. Sponsorship also makes it possible to create fictions[13] that associate the reputation of certain brands with the futuristic and science fiction imaginary world, which are considered virtuous by social categories with influence on the markets.

The institutionalization of science fiction, or *science fictionalization*, is the consequence of design fiction. If organizations decide to create their own science fiction, it is highly likely that this imaginary world will spread massively in the coming decades, as countries such as China develop these practices, aware of the advantage of these stories in order to innovate. Europe, the United States, Russia and Japan were already strong producers and consumers of science fiction. With design fiction, it is conceivable that the struggle for domination of the imaginary world will develop rapidly. By producing their own works of fiction, innovative institutions are taking storytelling to the next level. The imaginary world has succeeded in

13 The *Freedom* series (2006) is a Japanese serial manga that is part of a promotional operation called *Freedom Project*, launched by Nissin Foods on the occasion of its 35th anniversary. Sunrise Studios was chosen to produce a series of OVA (Original Video Animation) in which several scenes show characters eating instant noodles marketed by Nissin Foods. Product placement is a common practice in science fiction, with major brands seeing it as a way to reach opinion leaders.

The action takes place in the 23rd Century, more than a century after a catastrophe caused the collapse of Earth's civilization. A space station crashed on Earth, causing a series of cataclysms, wars and the disappearance of many forms of life. Humanity created a lunar city called Eden, capable of accommodating several million inhabitants, under domes reproducing the conditions of life on Earth.

channeling and modeling individual and collective dreams [PAR 01]. New works of fiction regularly emerge and are intended to achieve the status of sectoral myths, i.e. unifying narratives representing the objectives of an economic sector (e.g. terraforming for the space industry, or cyberspace for information technology) [KIN 00].

Organizations are engaged in a process of stimulating an internal positive imaginary world that contrasts with the external dystopian fashion [PAR 92, HIG 04]. The latter may be linked to a depressive mentality linked to the trauma of the September 11 attacks. However, large corporations, and especially the GAFAMs, have developed science fiction imaginary worlds and ideologies such as transhumanism to legitimize their existence in a technophile world. The technological utopianism of these actors contrasts with the fear of zombies and monsters that are increasingly present in the collective imagination and that bear witness to a macabre mentality symptomatic of collective fears, particularly collapsologists [SER 15]. Faced with the weakening of the utopian mentality, design fiction could revive positive thinking in organizations initially, and then in society, eager for ambitious collective projects to stimulate citizens, consumers and workers.

7.5. From the plausibility of design fiction to possible disappointment

Coulton *et al.* [COU 16] believe that design fiction can generate disappointment. A phenomenon of belief is indeed produced by this exercise of creativity, and may in some cases not produce the technologies and systems imagined. In particular, they suggest that:

> Plausibility is one of the key qualities often associated with suspension of disbelief, a quality encoded within the artefacts created as design fictions (…) designers may become minded to deliberately employ deceitful strategies in order to help their design fiction reach a larger audience.

The imaginary world can appear as a decoy, and considering it as a representation of a future in the making could be a strategic mistake for the R&D managers of a company or a State. Indeed, we should not take our own desires for realities, although such a position is tempting for the most innovative minds. Some science fiction stories project into the distant future, allowing us to envisage technologies that are far removed from the current possibilities of science. However, can we not consider the technological imagination as a prerequisite for scientific knowledge? [BRA 08] Before a major discovery, the imaginary world prepares consciences by representing the consequences of a possible paradigmatic disruption. The imaginary world and the symbolic thus ensure the passage from one reality to another. Supraluminic travel is an example of the technology that will be necessary for humanity if it wants to give itself the means to achieve its interplanetary and interstellar ambitions. This technology is a myth for the space sector [BOI 87], which for the moment has only engines that are unsuitable for remote exploration and exploitation of the cosmos, especially asteroids [DE 03]. The *Star Trek* franchise presents many utopian technologies, such as ships, capable of traveling at supraluminal speed. Many of the films inspired by this series use this mode of propulsion, for stories in which humans explore faraway lands [WES 00]. In this case, the imaginary world prepares minds in the long term for the development of this technology. Science fiction challenges researchers, and makes the public dream of a future multi-planetary civilization. It also warns against possible abuses of power by unscrupulous scientists. It is possible that major discoveries could lead to the development of a revolutionary propulsion system, which would quickly excite decision-makers and accelerate the conquest of the solar system. Science fiction participates in the construction of a collective, global imaginary world of the future of science and technology, presented as the vectors of transformation and social organization [BRO 09]. It is to be feared that this belief system, specific to the global technoscientific order, would collapse if civilization were to enter a phase of regression and if the population were to become impatient with the chronic impossibility of developing technologies that are nevertheless very realistic and plausible in fiction. Too much exposure to science fiction can thus lead to errors of judgment, exposing a scientist, for example, to confusion between the reality of knowledge in their field and a fictitious science that is nevertheless very realistic.

7.6. The theory of the failure of the imaginary world

Anthony Masure, professor of design, wrote an article [MAS 16] on the failure of the imaginary world, following Nicolas Nova's research, which states that "science fiction has lost its influence as a force of inspiration for the future"[14]. Worse, it would have the role of hindering the production of new imaginary worlds, generating a feeling of timelessness leading Bublex and During to consider that the future has disappeared [BUB 14]. Masure wondered how design could renew the imaginary world in order to act in the real world. He established a dialogue with the main arguments of Nova's theory. The latter stated that the role of science fiction is not to predict the future, or even to be useful. It should not be confused with futurology. Moreover, it is not intended to transform the world, but to entertain it. On the other hand, science fiction would have been overtaken by reality. Finally, science fiction is not about the future but about the present.

Designers are therefore not unanimous about the speculative and performative role of science fiction. This can be explained by the multiple forms of these stories. While some propose representational disruptions that are inspiring for innovators, many are of low quality and merely mediate theories that are already well known, with some simply repeating clichés that have already been assimilated many times by adapters of the genre. Design fiction could help reactivate futuristic imaginary worlds. If this method of stimulating creativity develops in the coming years, it is likely that new concepts will emerge from individuals and groups facing specific problems. Moreover, confronting new situations and seeking original solutions is the best way to generate new ideas. The science fiction dimension of design fiction is interesting because it refers to a successful fantasy that fascinates a large number of innovators. Creativity is stimulated by a playful and phantasmagorical approach referring to imaginary patterns shared by a global culture. Nova's arguments are interesting because the main themes of science fiction may seem dated, even outdated, to some observers. Other theorists consider that gender has become too dystopian, which could be detrimental to civilization, as the performative dimension of science fiction could lead some actors to realize negative scenarios [ALD 84]. For this reason, we have explained that institutions generally transform science fiction ideas before integrating them into their R&D programs and policy discourses [MIC 10a]. Design fiction is supposed to produce new imaginations. It will, however, be dependent on the global science fiction imagination, constituting a mythology for technoscientific societies.

14 Nova Nicholas, *Futurs? La panne des imaginaires technologiques*, Montélimar: Les moutons électriques, 2014, p. 5.

7.7. Science fiction prototyping and design fiction

Intel futurist Brian David Johnson recalls that most astronauts and cosmonauts were inspired by science fiction as children. He therefore proposes to extend this creative link to other economic sectors, such as artificial intelligence and computer science through science fiction stories. His aim is to develop the concept of "science fiction prototypes", which he believes are very influential in contemporary innovation processes. He quotes Bell and Dourish's [BEL 14] article, in which they explain how television programs such as *Dr. Who* and *Star Trek* are essential to the design of new technologies. Since then, the *Black Mirror* series has emerged and has had a significant impact on the representations of future technologies in the next 10–20 years. This program is particularly influential and mainly addresses changes in lifestyles under the impact of ICTs. Johnson states that:

> Science fiction does not merely anticipate but actively shapes technological futures through its effect on the collective imagination. At the same time, science fiction in popular culture provides a context in which new technological developments are understood. Science fiction visions appear as prototypes for technological environments[15].

He adds that "Science fiction is the playground of the intellect. Many people are mistaken when they think that the genre's aim is to predict the future (…) The future and science fiction have mingled together in our education and imaginations to such a point that there is no better medium to use as a platform for fictional prototyping"[16].

Johnson has developed alongside the work of Julian Bleecker, author of *Design Fiction: A Short Essay on Design, Science, Fact and Fiction,* a method of using science fiction to imagine possible worlds and make prototypes. Johnson's approach is particularly interesting because he applied it in the R&D laboratory of a large company. He even published a reference book [JOH 11], in which he explained his methodology illustrated with

15 Johnson Brian David, "Science Fiction Prototypes Or: How I Learned to Stop Worrying about the Future and Love Science Fiction", in Callaghan V., Kameas A., *Intelligent Environments 2009: Proceedings of the 5th International Environments*, Barcelona, 2009, p. 4.

16 Johnson Brian David, "Science Fiction Prototypes Or: How I Learned to Stop Worrying about the Future and Love Science Fiction", in Callaghan V., Kameas A., *Intelligent Environments 2009: Proceedings of the 5th International Environments*, Barcelona, 2009, p. 7.

numerous examples. Sterling, Bleecker and Johnson are three important authors in the history of design fiction. Subsequently, several authors interested in this concept have proposed original methods in articles, mostly applied to specific sectors such as energy and transport. Johnson's theories have had an impact mainly in the field of computer science and telecommunications, with Chinese trainers even using his approach to optimize the command of English and STEM in their universities. Microscience fiction prototyping is also being developed in parallel, consisting of writing tweet-length microfictions.

7.8. The pioneer, Julian Bleecker

Julian Bleecker is often cited along with Bruce Sterling as being one of the founders of design fiction [BLE 09]. His essay focuses on the use of this method to ensure the dialogue between science and science fiction. Design has the ability to propose prototypes that facilitate communication between the various disciplines in the innovation process. During his studies, Bleecker also worked on virtual reality, an area that is also in constant interaction with the imaginary world. He introduces the subject of his essay by seeking to abstract himself from the rigor of certain overly rational systems of innovation, his aim being to free and stimulate the imagination:

> Design fiction as I am discussing it here is a conflation of design, science fact, and science fiction. It is an amalgamation of practices that together bends the expectations as to what each does on its own and ties them together into something new. It is a way of materializing ideas and speculations without the pragmatic curtailing that often happens when dead weights are fastened to the imagination. [BLE 09, p. 6]

The author seeks to go beyond pragmatism and a form of rationality that are obstacles to creativity. Innovation processes are effectively limited by practical considerations, such as the search for budgets or the need not to produce inventions that could compete with products already on sale and economically profitable. Innovators have multiple mindsets that collaborate to generate new products. Design fiction participates upstream in crystallizing consumer expectations, but also and above all in designing prototypes capable of stimulating the imagination, scientists capable of making products out of them and consumers likely to buy them. Bleecker adds, "I began to wonder if science fact and science fiction are actually two approaches to accomplishing the same goal – two ways of materializing ideas and the imagination". From then on, he became aware of the possibility of using science fiction to develop new ideas and creativity in research projects: "How can science fiction be a purposeful, deliberate, direct participant in the practices of science fact?" [BLE 09, p. 6]. Bleecker believes that "Design fiction is a hybrid, hands-on practice that operates in a murky middle ground between ideas and their materialization, and between science fact and science fiction" [BLE 09, p. 9]. The author uses the concept of "diegetic prototype" to describe technical objects present in design fiction and science fiction. We have referred to these artifacts as "utopian technologies" in *Télécommunications et science-fiction* (2008). The concept of "diegetic prototype" is attributed to Kirby:

> I introduce the term diegetic *prototypes* to account for the ways in which cinematic depictions of future technologies demonstrate to large audiences a technology's need, viability and benevolence. Entertainment producers create diegetic prototypes by influencing dialogue, plot rationalizations, character interactions and narrative structure. These technologies only exist in the fictional world – what film scholars call the diegesis – but they exist as fully functioning objects in that world. The essay builds upon previous work on the notion of prototypes as *performative artefacts* [KIR 10].

We have established the performativity of certain science fiction works in previous studies, notably those devoted to the imaginary world of virtual reality [MIC 18]. To use the vocabulary of design fiction followers, it appears that the role of diegetic prototypes is to inspire innovators and guarantors of the strategic discourse of institutional leaders. Engineers and scientists must learn the design codes in order to optimize their creativity, and enrich their science fiction culture with new artistic concepts that go beyond works of global impact to propose fictional technologies in order to develop in the laboratory those that are likely to constitute marketable innovations. Managers, decision-makers and creatives participate in stimulating the imagination of artists and selecting the most promising works. The outline science–diegetic prototype–science fiction summarizes how design fits into the innovation process. Indeed, artists are playing an increasingly important role in the development of large companies' R&D strategies. It is no longer a question of envisaging the future, but of developing timeless concepts whose role is to explore and open up new avenues hitherto forgotten by scientists and engineers. An art form emanates from the technological culture of global capitalism. It emanates from the state of scientific knowledge and a science fiction imaginary world that needs to be surpassed and renewed. Indeed, design fiction develops concepts inspired by science and science fiction, while contributing to their renewal and dynamics. Thus, technological capitalism is developing an art form that is constantly changing, helping to capture consumer expectations and imagine a better world, because of commercialized innovations, sources of profits that also fuel R&D in which science fiction and design fiction artists rub shoulders with scientists and inventors.

7.9. Dreaming, a simulator of the dangers to come

Recent discoveries in the field of neuroscience have shown that dreaming has the role of simulating the potential dangers that the subject could face. For example, dreams of being attacked by wild animals are common among children, even in industrial cultures where this type of risk is unlikely. Anthropology has shown that there are imaginary archetypes common to all humans. Thus, dreams and the imaginary world can allow access to the unconscious, the center of these phenomena likely to structure and guide individuals in their confrontation with reality. The hypothesis of a dream being a simulator is interesting for our approach to the technical imagination. Science fiction is a form of fantasy that testifies to the role of the collective unconscious as theorized by Jung and later by many authors. We have theorized two specific forms of the unconscious, the prophetic unconscious and the technological unconscious, in *Télécommunications et science-fiction* [MIC 08]. The question of the origin of imaginary representations of utopian technologies was raised and it was established that the technical imagination drew its characteristics from the unconscious, endowed with a stock of images revealed to individuals according to their conditions of existence. The technological unconscious is thus the center of technotypes, technological archetypes whose theorization is of Jungian inspiration.

The utopian technologies of science fiction appear in an often dystopian or utopian narrative context. Critics of this artistic genre often denounce the violence and atrocities committed in certain films and novels. However, it is possible to see these stories as simulating the perils that industrial civilization will face. By anticipating the worst, science fiction writers prepare the public for disasters. For example, New York City has been destroyed hundreds of times in Hollywood movies, by natural cataclysms, alien invasions, wars and so on. Some critics regularly denounce the inability of the imaginary world to renew itself, to generate positive thinking. Anticipation and management of potential dangers could be functions of the unconscious mind. Imaginary machines are often the answer to dangers, such as mutant animal species, threatening enemies, or natural problems.

Dream theory, conceived as a danger simulator preparing individuals for potential dangers, is transferable on a community scale. The technological unconscious is thus punctually conscientized by the dream or the individual's imagination, before becoming a collective discourse following its formatting in the form of individual and then collective imaginaries. The transition from the dream to the collective imagination is a complex process that responds to the selection of the most relevant themes by companies according to their specific problems, often corresponding to strategic interests and identity.

7.10. Some approaches to design fiction

Some authors consider design fiction to be a pre-paradigmatic approach to innovation [COU 17]. The use of science fiction and the imaginary world is not new, but the development of pragmatic methods and the rationalization of creative processes have been helping to steer more and more research since the 1990s. Moreover, the number of academic articles devoted to design fiction is increasing every year. The imaginary world is less and less frightening for the rationalists in charge of managing the most innovative projects. Although science fiction is increasingly considered, it is still characterized by numerous scenes of violence that may put off some innovators. Design fiction allows us to get around the confusion between science fiction, violence and immorality. The works of fiction produced by this method are neutralized, i.e. they are shaped to match the strategic narratives of the companies and organizations that develop them. A company selling automobiles will be able to use fiction and prototypes without having to resort to a corpus that could scare off potential buyers. Design fiction is a clean type of science fiction, capable of imagining the future while remaining neutral and fitting into the ideological-strategic discourse of the organization that finances it. This practice gives organizations new power. They can create the fictions of the future, and disseminate them to the members of the structure, or even to the entire planet via the Internet, depending on whether the prototype, or fictional artifact, is to be kept secret or not. Internal design fiction can be used to spread a managerial vision. Clients and employees will be able to enter into a futuristic vision of the company because of these stories.

Science fiction offers stories that are sometimes considered to be industry myths. Robotics is influenced by Asimov's laws, virtual reality by the *Matrix* trilogy, and space travel by *Star Trek* and *Star Wars*, for example. These sectorial myths are generally created by artists outside the productive sphere, but generate collective representations of the use of highly influential futuristic technologies. Studying representations of the future of science fiction is therefore a necessity, as communities of engineers and researchers are often connoisseurs of the most innovative and influential films and novels in their field of investigation. Strategic intelligence allows us to consult and analyze new developments and to immerse ourselves in the history of science fiction, which often reveals utopian technologies that can potentially mobilize projects seeking to develop technological prototypes. Science fiction filmmakers often use designer artists to make credible the representations of the future in which they develop their stories. Science fiction is defined above all by the innovative technological and scientific elements found in the story. For example, Steven Spielberg used scientists to obtain their expertise in order to propose a credible futuristic setting for the film *Minority Report*.

Design fiction proceeds in a similar way. Experts in futurology, science fiction and many scientific and technical fields are called upon to create original works of fiction, the aim of which is to propel the organization to the pinnacle of creativity. A good work of fiction and a good prototype can mobilize more potential clients than a vision that is uninspired and not very credible. It is not easy to access certain design fictions, considered strategic by entrepreneurs. Others, on the contrary, see these creations as a means of publicizing the organization's activities. Design fiction, whether it is a short film or a science fiction anthology, can be used as advertising or even as a vehicle for corporate strategy, especially in the Internet age in which microdiscourse becomes viral and reaches potential clients, opinion leaders and influencers within hours. So far, few works of design fiction have become industry myths. It is possible, however, that some imaginary narratives have acquired the status of organizational technofiction. Some companies pay artists to help them create unifying works of fiction, some of which are in line with major sectorial myths, while others break away from all that is known and define new technological horizons through prototypes and revolutionary applications capable of guiding R&D and profits in the medium to long term. Fiction can therefore be considered as an advertising and managerial object, capable of motivating a population and structuring its vision of the future. It can also be kept secret, if it proves to be particularly relevant and innovative. Fiction therefore intervenes at different levels of the innovation process. It can stimulate the guarantors of the organizational narrative system (storytelling, sensemaking, design fiction). Researchers and actors in the field of fictionalization can meet from time to time in R&D centers to define the most relevant research avenues according to the scenarios imagined upstream. The imaginary world and science have an interest in merging from time to time to ensure a convergence of visions and optimize R&D. Fiction can also be crucial for the company's media coverage. A good design fiction can provoke a massive growth of potential consumers, who will recognize in the organization its pioneering dimension in the implementation of innovations that will inevitably be copied by competitors. Design fiction must therefore be used intelligently. The new projects, presented as confidential fiction during the time from prototype to marketable innovation, may be the subject of mass distribution for advertising purposes at a later stage. Design fiction stories are therefore at the center of an organizational narrative strategy whose role is to illustrate and reinforce the impact of management's strategic vision.

7.11. Science fiction, design fiction and foresight

Science fiction is often regarded as popular entertainment that is mainly distributed in industrialized societies [STA 07]. In recent years, it has become a means of anticipating, innovating and carrying out foresight in organizations, whether they are public institutions or large corporations. This approach has emerged in the United States, where science fiction is particularly legitimate, supported by a thriving cultural and film industry. It is part of the R&D programs of many institutions in Europe and around the world. This practice, which is increasingly widespread, contributes to constructing representations of the future that are both innovative and consensual on a global scale.

Stimulating the imagination and creativity has been a central concern of strategists since the 1950s. One of the first methods to carry out this important aspect of R&D was brainstorming. Creativity theorists have subsequently proposed dozens of methods promising the discovery of revolutionary solutions providing relevant answers to organizational problems and the invention of products or services that are sources of profit for companies. In the 1990s, design thinking appeared. It was a new approach to creativity that had a strong resonance. Coming from the United States, it was the subject of many books and management articles. At the center of its approach was the imagination of new solutions, through the process of emergence of new ideas, called ideation.

At the same time, science fiction developed at a rapid pace, with Hollywood blockbusters reaching a global audience. With the emergence of *Star Wars* in 1977, science fiction reached a new scale and reached a large audience. Special effects enabled futuristic technologies to be represented, due in part to the inventiveness of brilliant designers. The new science fiction films all featured utopian technologies, which later inspired engineers and scientists to put their knowledge to work in R&D for innovations that could change the world and bring in billions of dollars. From then on, innovation, which, together with design thinking, set itself the goal of imagining and designing prototypes, was interested in the science fiction imaginary world.

Despite the mockery and reticence of some actors who believed that science fiction was a metaphor for the present and not a futuristic projection, other, more pragmatic actors saw it as a source of new ideas, or even the imaginary fuel necessary for a technological capitalism with growing influence on the global productive system.

A detour through the history of science fiction and the history of technology shows indeed that this imaginary world has often been at the forefront of major innovations since its invention in the 19th Century. Over the years, this entertainment for the popular masses has thus become a means for artists to propose utopian technologies inspired by science and at the origin of new theories and new technoscientific discoveries. Telecommunications, space, but also the medical and energy sectors were at the center of this technical imagination.

However, research centers have set out to map this imaginary world, with the aim of anticipating technological changes and possibly funding projects that could make science fiction come true. While NASA and ESA took an early interest in these ideas, even organizing science fiction competitions to probe the collective imagination, GAFAM, which is said to realize the cyberpunk imaginary world, has on several occasions even proposed its own science fiction stories, publishing anthologies looking at the future of economic sectors such as IT, health, nanotechnology and biotechnology.

Design fiction is being used more and more, especially in Europe, and is a continuation of design thinking [MIN 16]. The aim is to speculate on the future with prototypes and fictions that envisage the imaginary transformations of a sector. After detecting weak signals and problematizing a situation, innovators move on to creating new ideas and concepts. The imaginary world is at the center of their interests, because it is thanks to it that engineers and scientists will work in a direction that will be a carrier and source of new technologies potentially generating profitability and profit. Fiction is a way to stimulate the imagination. The followers of design fiction develop in organizations' practices of creating technical or even science fiction. The most innovative concepts can then be prototyped, which strategists can in turn integrate into their innovative discourses and activities.

The practice is increasingly developing in a knowledge-based economy that is adept at such approaches. Even the French army, in 2019, decided to use science fiction authors to help strategists define concepts on the future of war and consider the most dangerous situations for national security.

Science fiction and innovation are therefore at the center of global capitalism. While some actors frequently believe that science fiction does not have the function of predicting the future, or that this fantasy is already outdated, the facts point in the opposite direction. Fiction, whether written or in video form, contributes to corporate identity. Creating organizational technofictions could well be a new fashion contributing to the construction of corporate identity. Design fiction thus ensures the fusion and the overcoming of design thinking and storytelling, contributing to creativity and the development of foresight [MIC10a].

7.12. Toward a new mythology because of storytelling

The technical imagination is thus based on utopian technologies at the center of science fiction stories, which popularize the research being carried out in the laboratories of their production companies. Contemporary myths are heavily influenced by science fiction and fantasy, the former generally taking place in the future while the latter take place in worlds inspired by medieval culture. The imaginations of "geeks" are thus particularly flourishing and influential, launching from California mythologies with a universal vocation, elaborated by experts in storytelling and cultural soft power. The Americans have developed expertise in writing planetary mythology, likely to take over from cultural systems that have become inoperative. The consequence of capitalism and innovation is exhausting cultural systems to the point of making them obsolete, thus threatening the social stability of some societies, and frequently the explosion of community systems. The engineers of the global imagination are inspired, for example, by the methodology of Joseph Campbell, who, through his monomyth theory, reveals the universal discursive structure of myths [CAM 49]. Thus, by better understanding how mythology functions on several continents, it becomes possible to model the imaginary world with a view to creating contemporary myths, adapted to the evolution of society and the productive system. Capitalism thus regularly renews the imagination that legitimizes it and popularizes its new flagship technologies. Science fiction plays the role of an advertising imaginary world for the innovations structuring the new economic cycles. The imaginary world can be analyzed under two main aspects.

On the one hand, it is said to be at the origin of inventors' creativity. Engineers are influenced by science fiction to direct their research and gain popular support. For technoscience to be successfully sustained in the world, it must meet expectations, a technical imagination that has been promising the earth for several decades [HOT 13]. Science fiction can be analyzed as an element of the scientific imagination, which it helps to structure and format in the form of imaginary worlds, thus making it possible to direct this research. At the same time, it structures consumer expectations.

On the other hand, it can be seen as part of the ideological apparatus of technocapitalism, conscious of the need to raise expectations by spreading a mobilizing phantasmagorical imagination. Capitalism has no particular religion. On the other hand, it has an increasingly flourishing imagination because it serves to structure a collective, global mentality with each new economic cycle [NOB 97]. Innovations thus have more difficulty spreading throughout the world if the production system is not equipped with a soft power unifying consciences around technical imaginations playing the role of real sectorial myths.

Virtual reality researchers have been inspired by novels and films from the 1980s onwards, which have evolved from a subculture to a global mythology [KER 07]. Science fiction writers have sometimes been criticized for not anticipating the Internet, proving that the theory of the prophetic function of this futuristic imaginary world was wrong. We agree that this prophetic function would be exaggerated. However, this imaginary world has been playing a contributory role to the innovation process since the 1980s, representing the immersive virtual worlds of the future, which will make the Internet look like an antiquity in the coming decades. Science fiction may not have anticipated the Internet, although some stories prior to this invention have described the interconnection of computers, but it does contribute to the preconception of a global immersive system in which people will spend much of their time when innovations are mature and commercialized [BUK 93].

Moreover, it would be an exaggeration to consider that this imaginary world occupies a place similar to that of Protestant ethics in the spirit of capitalism theorized by Max Weber [WEB 98]. The latter had also shown the impact of Judaism on the economic sphere and for the elaboration of certain prophecies. Religion plays an important role in social stability and the justification of the economic system. It also creates cultural differences that can hinder the development of capitalism, which must in response build its own spiritual, imaginary system, in order to structure a technical culture at the origin of the creativity of inventors and the consumerist desire of the world's population [BIL 14].

7.13. From utopian technologies to the technotype theory

The utopian technologies of science fiction can therefore be used to activate an innovative mentality. The imagination of designers, engineers and scientists is stimulated by these artifacts. Part of the act of invention has a creative dimension, and imagination plays an important role in the development of new scientific theories and technologies. Technical innovation is thus based on an imaginary system of varying degrees of awareness [MAL 00] that companies and organizations are trying to better understand in order to integrate it into their research and development strategies [WEI 95].

Are utopian technologies individual creations, or are they to be found in what we propose to name after Jung [JUN 90] a technological collective unconscious? Should we prefer the term collective technical imagination, or collective representations of technology? The choice of terminology is important, as these three expressions refer to different and complementary states of human creativity.

The technician collective unconscious groups together technotypes. Jung explains that the archetypes of human psychology are found in the collective unconscious. The technologies developed by the inventors would be in this unconscious, latent state, before being discovered as innovations [GOL 95].

Technotypes, like identity archetypes, belong to an ancestral psychology. Their study provides a better understanding of how the economy works, as innovations are linked to an analysis and discovery of these technotypes. The inventor proofreads one or more technotypes. He updates these unconscious structures in the light of his technical knowledge and develops new theories and concepts. According to this approach inspired by psychoanalysis, there are individual and collective technotypes. The theory of the unconscious mind was put forward by Freud at the individual level, before being extended to the collective by Jung. The latter finds, particularly in myths and tales [BET 76], expressions of the collective unconscious. Our science fiction research proceeds in the same way. Films and novels would testify to a collective unconscious. The technologies included are technotypes. The theory of the technician collective unconscious considers that a spiritual world preexists at the collective level and feeds the history of humanity. However, it is necessary to consider that inventors proceed according to an approach based on experience and individual deduction from experience. In this way, their imagination meets the technotypes, which provide access to scientific and technical knowledge common to all humans. Through their research, scientists are contributing to a better understanding of human technotypes. Interindividual exchanges allow the enrichment of the technological collective unconscious. This unconscious can, however, be better known through the study of its expression, the technical imagination and utopian technologies.

The collective technical imagination has developed mainly since the industrial revolution through art and science fiction. A transformation took place from the 17th Century onwards, to the point of pushing the philosophers of the Enlightenment to conceive the mission of human civilization to master nature through science and technology [ALK 02]. From then on, an abundant technical imagination developed in Europe and America, activating a collective innovation process with uncontrollable consequences. The technical imagination was the consequence and the cause of many discoveries arising from the scientific method [FLI 01]. While the technical imagination was so successful, it was because it referred to research programs inaccessible to laypeople [GRA 13]. Science fiction became a popular culture, capable of propagating and popularizing the most complex theories and investigations. A collective technical imagination developed in the West, to the point of becoming, after the Second World War, a tool for questioning science in ethical terms, and for illustrating the most radical theories on invention, and above all on the dissemination of research projects in search of funding. Science fiction allowed the creation of a global technical imagination, making popular a science that for many remained abstract and coldly instrumental, far removed from the daily concerns of a large part of the population [WAR 82]. Because of science fiction, innovation became popular. However, the representations of scholars were the subject of caricatures and social criticism by many authors. The collective imagination helped to represent the scientist as an unscrupulous being, whose power had to be controlled in order to avoid unethical or even criminal experiments. The archetype of the mad scientist follows in the footsteps of the seminal work *Frankenstein,* who eventually created a monster potentially dangerous to the survival of the human species. The technical imagination is ambivalent, spreading utopian, then dystopian, often critical and metaphorical accounts of the state of science at a given time. This imagination, although critical, shaped at different times the collective representations of a potentially Promethean science. The so-called "geek" generation was thus rocked by a "technoid" science fiction coming mainly from the United States and Japan, the most innovative countries in the years 1980–2010. Different media presented utopian technologies, such as flying cars, spaceships, robots or virtual reality, as objects for common use in the medium term. The technical imagination is a pioneer in the innovation process [EAS 08]. It is most often found in a futuristic environment, and raises the question of the interest or dangers of certain inventions. Thus, a technology can be considered virtuous if it is controlled by positive individuals. Conversely, it can become dangerous if negative beings get hold of it.

Comics, pulps, then television and the Internet made it possible to broadcast many works of fiction, shaped to ensure global distribution. The inhabitants of the most developed countries can thus have access to a collective technical imagination, to the point of developing common representations of the technological future. In this way, the whole planet participates in innovation, following images of technologies to be developed. The technical imagination is important because it allows the development of a consensus around medium- and even long-term horizons. The R&D of global capitalism thus seeks to realize utopian technologies that can be considered as advertising images of the productive system.

Collective representations of technology refer to Durkheimian terminology. Myths allowed primitive societies to build collective beliefs. These representations allow a group to unify to generate social bonds and to project itself toward the future. Durkheim states that the ability to create collective representations distinguishes humans from animals and helps to structure societies [DUR 15]. Religions and myths are examples of belief systems that have been used to construct complex societies [DUR 00]. In this context, technoscientific societies also need collective representations that can unify and unite them. A strong social bond allows a group to increase its performance and become more powerful. Collective representations of technology are thus produced by an imaginary world of which science fiction is a part. We found that technotypes and then technical imagination helped to stimulate scientific creativity [IYA 83]. The individual, in his research activity, must rely on his objective knowledge and his subjectivity as a discoverer. He must also be inspired by collective representations in order to become one with his society. The collective unconscious houses technotypes of which knowledge is still partial. However, it manifests itself in the form of myths [ANT 85], some of which end up being made conscious to the point of becoming collective representations, or even popular beliefs sometimes ritualized within religious systems. With science fiction, the global technoscientific society develops its collective representations. The great innovative trajectories are admitted, and the most influential technotypes are listed, analyzed and enriched by an imaginary industry whose aim is to shape the human imagination into being performative and socializing, through what Americans sometimes call soft power.

R&D is developing at a global scale, with most major innovations being generated by networked and interconnected laboratories [LAT 89]. In this context of a globalization of research, the common imagination can be considered as a vector of cohesion and collective stimulation around common interests. Moreover, the Internet has made it possible to optimize overall R&D by facilitating access to knowledge and projects that would have been difficult to consult previously. Thus, since the corpus of works of fiction constituting the collective representations of technology is limited, it is possible to consider that the collective imagination has finally reached the limits of its beneficial effects for innovation. By dint of referring to the same works, innovation stagnates and creativity is no longer stimulated as much. It is therefore necessary to develop a new imaginary system, allowing capitalism to be relaunched into new economic cycles. Economic cycles fluctuate according to the imagination that accompanies them. Imaginations are prepared before contributing to the invention and then to the diffusion of an innovation. Then they become less and less efficient, which also translates into the end of the technology cycle they were supporting. Other imaginary systems, which developed in alternative systems, take over by promoting new technologies at the origin of economic cycles that ensure the perpetuation of capitalism. Technological imaginary worlds evolve and transform according to the place of the invention they represent in the economic system. Sometimes emerging, sometimes dominant and sometimes ineffective, they are symptoms of the success or failure of innovations. At the end of a cycle, it is therefore necessary to relaunch the search for imaginary ideas that could potentially structure the productive system's members. Researchers, salespeople, strategists, decision-makers, administrative agents, consumers, etc., all must adhere to a system of beliefs and imaginations that can integrate innovation into society.

Thus, the global innovation pact may well have to adapt to new imaginary worlds if its constituent technologies become obsolete or too difficult to commercialize in the medium term. This is why companies need to distance themselves from the global imaginary world in order to conceive their own, as far as possible breaking with already obsolete representations. Inventing the new economic cycle requires the development of a critical imagination. Science fiction can thus be analyzed as a genre that is very critical of technostructure. It is not uncommon for these stories to denounce the negative impact of technoscience on the individual, society and nature [MOY 00]. Businesses and organizations are therefore tempted to detach themselves from this type if it becomes too critical of their economic sector. They then develop their own fictions, inspired by existing and reformatted works, or by appealing to creative minds, to artists like science fiction writers. Imagination specialists are thus used to train employees in creative methods such as brainstorming, design thinking or design fiction. A multitude of micro-imaginations are emerging. The individual imagination of anonymous employees is then valued, as it is likely to confer originality on the organization. Individual imaginations are valued, protected and integrated into the innovation process. The thousands of imaginations emanating from these sessions are supposed to feed the researchers likely to realize the utopian technologies. Technofiction is a preshaped fantasy created with no commercialization goal, only to serve as a focus for the organization's strategists and innovation scientists. It may also be used in advertising campaigns if the technology it describes becomes a commercialized innovation. The multiplication of science fiction is the result of a productive process leading to the inefficiency of obsolete dominant imaginary worlds. A new economic cycle is the product of this new creativity, which generally thwarts the predictions emanating from the analysis of the most obvious imaginings and trends. The study of weak signals makes it possible to capture the content of a few technofictions, the most effective of which will be those that have enabled the development of inventions patented and marketed by a limited number of players. The multiplication of technofictions after the end of an economic cycle makes it possible to revive competition and rivalry between the actors of the productive system. The struggle to impose a technical imagination through innovation is then accompanied by competitive processes leading to the disappearance of certain players and the emergence of entrepreneurs with the most original and performative imagination.

7.14. Four proposals on technotypes

Thus, the emergence and functioning of technotypes responds to a few fundamental stages. They are the result of a recollection of ancestral events and the activation of fundamental archetypes:

1) Archetypes are the product of humanity's adaptation to the difficulty of surviving in the natural state.

2) As humanity has emerged from the state of nature, it has gradually generated imaginary archetypes, products of the archaic imagination. Technotypes are specific archetypes, relating to technology, peculiar to humankind (with a few exceptions for apes).

3) Gradually humanity developed a technosphere, particularly important and influential with the industrial revolution. The technical imagination comes from technotypes. Scientific discoveries rationalize the natural environment, but also reveal new dangers such as pollution, the arrival of an asteroid and the cosmic immensity, exciting the imagination, which proposes new technical solutions.

4) The archaic imagination, particularly in relation to humanity's place in the universe, is at the origin of religions, but also of technotypes at the origin of the technical imagination. Art, since prehistoric times, shows the

usefulness of technology in taming nature, as in hunting scenes. The archaic imagination has its roots in an ancestral memory, but also constitutes a fundamental mental structure. The process of passing from the imagination to the imaginary world is caused by a phenomenon of rationalization that contributed to the birth of civilizations, the most powerful of which were those that mastered the dominant technologies, particularly military ones.

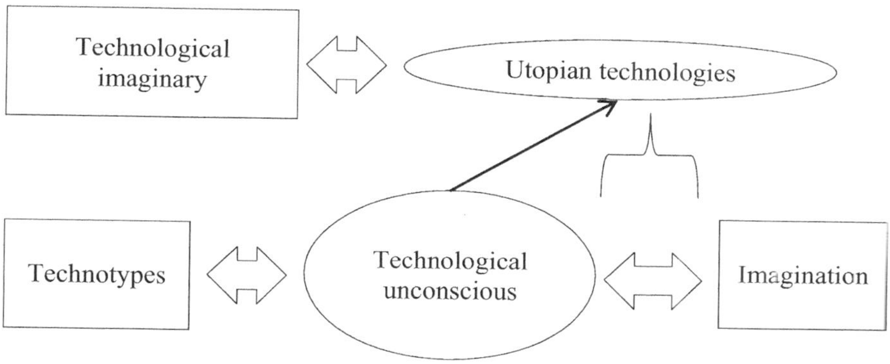

Figure 7.1. *Technotypes and utopian technologies*

7.15. Beliefs and behavioral economics

The economy is driven by imaginary flows, in turn structuring technological ideologies, one of the most widespread avatars of which is transhumanism. Economic actors are supposed to make their decisions and act according to the laws of the market in the case of the liberals, and according to the class struggle in the case of the Marxists. However, both paradigms are based on adherence to an imaginary that structures competition and power relations. The behavior of the economic actor varies according to an imaginary that is disseminated in his production sector. Financial market participants are not immune to the influence of the imagination. Decisions are therefore not purely rational, despite the predominance of science and mathematics in contemporary economics. In order to innovate, players must anticipate demand and prepare products in R&D laboratories that come from the imagination of individuals or teams of researchers. The customers' imaginations, if it is partly shaped by marketing, reflecting an organizational imagination, also influence entrepreneurs. Science fiction intervenes at this level at the interface between the client and the innovator. A shared imagination ensures the success of the innovator and the satisfaction of the buyers. A pathological imagination can lead to rejection and the failure of a product, particularly through its moral condemnation. It remains to be determined how the imagination of traders and financiers works in order to envisage its impact in crises such as that of 2008. Is an economic crisis linked to the spread of an imaginary that is incompatible with the functioning of capitalism, or the manifestation of a change of economic paradigm, structured in part by imaginary worlds that have become obsolete and are being replaced by new systems of representation?

7.16. Realistic, imaginary systems and their cyclicity

The relationship between imagination and reality is both the cause and the consequence of historical dynamics. In an individual, rationality and imagination are faculties that go hand in hand, one dominating the other according to the normative system governing the social order.

– An exacerbated rationality in a society whose organization is based on fictions such as religions would risk excluding the individual. Think of the example of Galileo Galilei who withdrew his astronomical theory to avoid being condemned to death by the Church, which threatened to denounce him as a heretic. In many contemporary dictatorships, for example in some Muslim states, rational and atheistic thinking is considered suspicious and even unacceptable. Theocracies are highly imaginative systems, with institutional works of fiction being presented as truths superior to scientific considerations.

– An overdeveloped imagination in a realistic society, for example technoscientific, could be problematic, as individuals developing them are considered psychologically unsuitable, or not very serious, or even infantile. Technocracies, for example, hunt down subversive elements, which may present the imaginary as a means of fighting against an all-powerful instrumental rationality, leading for example to the dehumanization of the system. The call for revolt through the imagination is recurrent in many social or revolutionary movements.

Finding a balance between the imaginary world and reality is a necessity in a social or organizational system. If the individual has to adapt to the dominant norm, whether it be realistic or "imaginist", it is also up to society to organize itself to make possible the expression of discourse of all kinds, whether imaginary or scientific. Moreover, most of the great thinkers of the imagination consider the communication between these two psychic faculties as a necessity. Permanently atrophying one of the two can only be a source of disorder in the medium term, as the impossible expression of the imaginary world, of rational thought, or even of a symbolic dimension risks leading to a discursive repression that is harmful to the stability of organizational systems.

The history of human civilizations shows the existence of imaginary cycles, to be compared with the expression of realism. Since the beginning of the industrial revolution, this cyclicity has been particularly evident. Religions are contested by a scientific realism which, however, elaborates a rich foundation of fictions and fantasies, whose function is to ensure the stability of the social order and to motivate individuals participating in the economic system. The great discourses have regressed but have given way to micro-narratives in permanent renewal. Triumphant science has spread in the West and Asia in a spectacular way since the beginning of the industrial revolution to the point of ensuring the development of an imagination adapted to its ambitions and powerful enough to legitimize the perpetuation of the technoscientific structure on a global scale.

7.17. Conclusion

It has become common to consider science fiction as a particularly influential culture in the construction of collective representations of the future. It is not surprising, therefore, that this popular culture is scrutinized with an interested eye by the actors responsible for producing successful innovations. Researchers in R&D centers are making every effort to detect underground trends, with weak signals announcing the most radical changes in society almost imperceptibly. Predicting the future is the ambition of innovators, who wish to direct their R&D budgets toward particularly promising futuristic technologies. If people increasingly feel that they live in a world of science fiction, it is because the technological sectors are driven by a collective dynamic generating innovation at an ever faster rate, plunging a large part of humanity into an abundance of inventions. Product life cycles are getting shorter, and modes are changing at an increasingly impressive rate. The technoscientific era feeds on a rich imagination, which in turn draws inspiration from innovations to create futuristic worlds with utopian technologies. In the medium term, science fiction is about to change.

It will take the form of design fiction, a creative technique derived from design thinking and storytelling, and promises to multiply science fiction stories exponentially. Indeed, companies, public institutions, associations and all other forms of organizations are trying to describe their future to make it more innovative, but also more controllable and manageable. Paradoxically, design fiction should make it possible to innovate within a shaped imaginary framework. By helping actors create a futuristic scenario, design fiction contributes to their identity construction. Science fiction identities are fashionable in societies where this genre has invaded the screens since the 1970s. With design fiction, a flood of futuristic representations is announced. Science fiction should undergo change, even if the question of its identity is raised. Indeed, the stories of design fiction, which we also call institutional science fiction, will in most cases be inspired by founding narratives that have marked the collective imagination, sometimes for decades. Inventing a utopian technology specific to a sector of activity is often the objective of the creative process. The actors dominating the imaginary world are presented as the future leaders, provided they can transform fiction into reality, because of the process of technoscientific innovation. A new theory or patent may be the subject of fiction, envisaging its use and social application. Every sector has its founding myth, a technical imagination that is often performative. Innovators believe in a future and eventually realize it, because of their knowledge and R&D. So will design fiction succeed in freeing itself from the history of technology in order to propose utopian technologies that are in total rupture with representations of the past? It would be pointless to try to break with the founding narratives of the genre. On the contrary, creativity is often the product of many cultural strata. It is therefore important, even necessary, to have recourse to the technical imagination of a sector before envisaging the creation of new stories. Design fiction studies the history of science fiction before launching the creative process and the search for new subjects of fiction. Thus, including a creative process in a history of imaginary techniques makes it possible to inscribe creativity in a technoscientific lineage. Institutions are also the product of history, and it is important to remember how the technologies that have made them successful in the past have also helped to invent new ones.

Writing fiction has become a widespread practice for technological foresight, but also for writing in the humanities. An imaginary story helps to illustrate new approaches. Fiction is thus seen as an element of rationalization.

However, it remains difficult to access technotypes, and the study of the technical imagination is a discipline that requires recourse to what Gilbert Durand called archetypology [DUR 93]. Similarly, if the fundamental technotypes are conceived as the generators of the utopian technologies guiding the technoscientific system, it is not absolutely necessary to consider the use of cognitive science or brain imaging to locate them in the brain. The utopian technologies present in science fiction stories are manifestations of a technological unconsciousness harboring these technotypes. Therefore, the question is to determine how the process of generating utopian technologies by technotypes works. This fundamental research also raises the question of the conditions for the emergence of new myths in a culture. The mythological imagination, like the technical imagination, activates archetypes (and technotypes), with a view to creating new images. Thus, cultural filtering operates to innovate and create utopian technologies. A society will, for example, be faced with new livelihood conditions and will develop science that can help it to cope with this situation. The technological unconsciousness is then activated and stimulates the technotypes likely to generate utopian technologies adapted to the case to be treated. Culture then intervenes to formalize the production of the technological unconscious into a technical imagination. From then on, an innovation process takes place, the culturally formatted imagination being endowed with performative power.

Does this mean that the nature of technotypes is universal? This question refers to the hypothesis of a universal collective unconscious developed by Jung. Other thinkers, such as Vernadski and Teilhard de Chardin, have evoked the existence of a noosphere, welcoming the imagination, myths, ideologies, in short, human thought. The question of noogenesis and technotypic universality remains. Science fiction is particularly well developed in the most developed, industrialized countries. The technical imagination is evolving, particularly in the most intense areas in terms of technoscientific research, such as the United States, Europe and Japan. Other territories, less developed, are in a theocratic era. Finally, the least industrialized areas remain attached to a mythical, prereligious mentality. These three categories probably all have access to a technological unconsciousness, more or less stimulated, due to a variable interest in technology. The most developed societies, however, use science fiction as a mythology of the technoscientific era. Ancient myths also dealt with technology, such as the characters Prometheus and Icarus. The technical imagination has existed for a long time in the history of humankind. It develops particularly in periods of great discoveries, such as the European

Renaissance, the Industrial Revolution and the Technoscientific Era. A paradox appears. Myths are supposed to provide social stability by constituting immutable collective beliefs. With science fiction, we are witnessing a permanent movement. Innovation is at the heart of the economy, and societies are forced to constantly adapt to new cycles that have emerged as a result of the takeover of new technologies. Mythological production is thus created by a social need for stability. Imagination and reality are in constant interaction. Thus, the function of utopian technologies and technotypes in collective cognition is to provide images of machines capable of modifying and adapting society to new situations. From then on, science fiction mediates between the imaginary world and science to the point where it is considered to be an essential element of R&D in the technoscientific era.

7.18. References

[ALT 96] ALTER N., *Sociologie de l'entreprise et de l'innovation*, PUF, Paris, 1996.

[ALD 84] ALDRIDGE A., *The Scientific World View in Dystopia*, UMI Research Press, Ann Arbor, MI, 1984.

[ALK 02] ALKON P.K., *Science Fiction Before 1900: Imagination Discovers Technology,* Routledge, London, 2002.

[ALK 10] ALKON P.K., *Origins of Futuristic Fiction*, University of Georgia Press, Athens, 2010.

[ALT 64] ALTSHULLER G.S., *How to Learn How to Invent*, Tambov Publishing House, Tambov, 1964.

[ALT 84] ALTSHULLER G.S., *Creativity is an Exact Science: The Theory of the Solution of Inventive Problems*, Gordon and Breach Science Publishers, London, 1984.

[ALT 07] ALTSHULLER G.S., The Innovation Algorithm, TRIZ, Systematic Innovation and Technical Creativity, Technical Innovation Center, INC, Worcester, MA, 2007.

[ALT 10] ALTER N., *L'innovation ordinaire*, PUF, Paris, 2010.

[AMA 83] AMABILE T.M., *The Social Psychology of Creativity*, Springer-Verlag, New York, 1983.

[AMA 96] AMABILE T.M., *Creativity and Innovation in Organizations*, Harvard Business School, Boston, MA, 1996.

[ANT 85] ANTCZACK J., *Science Fiction: The Mythos of a New Romance*, Neal-Schuman, New York, 1985.

[ARO 95] ARONOWITZ S., MARTINSONS B., MENSER M. *et al.* (eds), *Technoscience and Cyberculture*, Routledge, London, 1995.

[BAC 87] BACHELARD G., *L'eau et les rêves: essai sur l'imagination de la matière*, J. Corti, Paris, 1987.

[BAC 87] BACHELARD G., *Le nouvel esprit scientifique*, PUF, Paris, 1987.

[BAC 90] BACHELARD G., *L'air et les songes: essai sur l'imagination du mouvement*, J. Corti, Paris, 1990.

[BAC 92] BACHELARD G., *La poétique de l'espace*, PUF, Paris, 1992.

[BAC 00] BACON-SMITH C., *Science Fiction Culture*, University of Pennsylvania Press, Philadelphia, PA, 2000.

[BAI 86] BAINBRIDGE W.S., *Dimensions of Science Fiction*, Harvard University Press, Cambridge, MA, 1986.

[BAL 01] BALANDIER G., *Le grand système*, Fayard, Paris, 2001.

[BAR 07] BARBROOK R., *Imaginary Futures – From Thinking Machines to the Global Village*, Pluto Press, London, 2007.

[BAU 98] BAUDRILLARD J., *Amérique*, Le livre de poche, Paris, 1998.

[BEL 14] BELL G., DOURISH P., "Resistance is futile: reading science fiction alongside ubiquitous computing", *Personal and Ubiquitous Computing*, vol. 18, no. 4, pp. 769–778, 2014.

[BER 96] BERGER R., *L'origine du futur*, Ed. du Rocher, Monaco, 1996.

[BER 38] BERGSON H., *La pensée et le mouvant*, PUF, Paris, 1938.

[BER 86] BERNABEU E.P., "La science-fiction: une nouvelle mythologie", in *Science-fiction et psychanalyse, l'imaginaire social de la S.F.*, Bordas, Paris, pp. 222–229, 1986.

[BET 76] BETTELHEIM B., *Psychanalyse des contes de fées*, Robert Laffont, Paris, 1976.

[BIL 14] BILTON C., CUMMING S., *Handbook of Management and Creativity*, Edward Elgar, Cheltenham, 2014.

[BLE 09] BLEECKER J., Design Fiction: A Short Essay on Design, Science, Fact and Fiction, Near Future Laboratory, San Francisco, 2009.

[BLY 05] BLY R., *The Science in Science Fiction: 83 SF Predictions that Became Scientific Realities*, BenBella Books, Dallas, 2005.

[BLY 14] BLYTHE M., "Research through design fiction", *Proceedings of the 32nd Annual ACM Conference on Human Factors in Computing Systems*, CHI '14, ACM Press, New York, pp. 703–712, 2014.

[BOI 87] BOIA L., *L'Exploration imaginaire de l'espace*, La Découverte, Paris, 1987.

[BOI 89] BOIA L., *La Fin du monde, une histoire sans fin*, La Découverte, Paris, 1989.

[BOI 98] BOIA L., *Pour une histoire de l'imaginaire*, Les Belles Lettres, Paris, 1998.

[BOL 99] BOLTANSKI L., CHIAPELLO E., *Le nouvel esprit du capitalisme*, Gallimard, Paris, 1999.

[BOO 06] BOOKER M.K., *Alternate Americas: Science Fiction Film and American Culture*, Greenwood Publishing Group, Westport, 2006.

[BRA 08] BRAKE M., HOOK N., *Different Engines – How Science Drives Fiction and Fiction Drives Science*, Macmillan Science, New York, 2008.

[BRO 09] BROWN T., *Change by Design: How Design Thinking Transforms Organizations and Inspires Innovation*, Harper Bus, New York, 2009.

[BUK 93] BUKATMAN S., *Terminal Identity: The Virtual Subject in Postmodern Science Fiction*, Durham, Duke University Press, London, 1993.

[BUB 14] BUBLEX A., DURING E., *Le futur n'existe pas: rétrotypes*, B42, Paris, 2014.

[BUD 00] BUDERI R., *Engines of Tomorrow: How the World's Best Companies are Using Their Research Labs to Win the Future*, Simon & Schuster, New York, 2000.

[CAM 49] CAMPBELL J., *The Hero with a Thousand Faces*, Pantheon Books, New York , 1949.

[CAN 93] CANTO C., *The History of the Future: Images of the 21st Century*, Flammarion, Paris, 1993.

[CAR 06] CARRET P.L., *The Art of Speculation*, Cosimo, New York, 2006.

[CAR 11] CARRIER C., GELINAS S., *Créativité et gestion, les idées au service de l'innovation*, Presses de l'Université du Québec, Quebec, 2011.

[CAS 99] CASTORIADIS C., *L'institution imaginaire de la société*, Le Seuil, Paris, 1999.

[CAV 01] CAVALLARO D., *Cyberpunk and Cyberculture: Science Fiction and the Work of William Gibson*, The Athlone Press, London, 2001.

[CAV 16] CAVALUCCI D., "Genrich Altshuller. Théoriser l'acte inventif pour mieux l'enseigner", in BURGER-HELMECHEN T., HUSSLER C., COHENDET P. (eds), *Les grands auteurs en management de l'innovation et de la créativité*, Editions EMS, Paris, 2016.

[CAZ 86] CAZES B., *Histoire des futurs, les figures de l'avenir de Saint-Augustin au XXIème siècle*, Seghers, Paris, 1986.

[CHA 13] CHAPMAN J., CULL N.J., *Projecting Tomorrow: Science Fiction and Popular Cinema*, I.B. Tauris, London, 2013.

[CHO 15] CHOUTEAU M., FAUCHEUX M., NGUYEN C., "Les élèves ingénieurs à l'épreuve de la fiction. Quelles relations entretiennent-ils avec les mondes scientifiques et techniques fictionnels?", *Les Enjeux de l'information et de la communication*, vol. 16/3A, 2015. http://lesenjeux.u-grenoble3.fr/.

[CLA 05] CLAMEN M., *Jules Verne et les sciences cent ans après*, Belin, Paris, 2005.

[CLE 15] CLEGG B., *Ten Billion Tomorrows: How Science Fiction Technology Became Reality and Shapes the Future*, St. Martin's Press, New York, 2015.

[COU 16] COULTON P., LINDLEY J., AKMAL H.A., "Design fiction: does the search for plausibility lead to deception?", *Proceedings of Design Research Society Conference 2016*, Design Research Society, GBR, pp. 369–384, 2016.

[CSI 90] CSIKSZENTMIHALYI M., *La créativité: psychologie de la découverte et de l'invention*, Robert Laffont, Paris, 1990.

[DE 03] DE WITT D.K., *Astrofuturism: Science, Race and Visions of Utopia in Space*, University of Pennsylvania Press, Philadelphia, 2003.

[DIS 00] DISCH T., *The Dreams Our Stuff is Made Of: How Science Fiction Conquered the World*, Free Press, New York, 2000.

[DOS 82] DOSI G., "Technological paradigms and technological trajectories: a suggested interpretation of the determinant and direction of technological change", *Research Policy*, vol. 11, no. 3, pp. 147–162, 1982.

[DRE 06] DREGNI E., DREGNI J., *Follies of Science – 20th Century Visions of Our Fantastic Future*, Speck Press, Denver, 2006.

[DUN 13] DUNNE A., RABY F., *Speculative Everything: Design, Fiction, and Social Dreaming*, MIT Press, London, 2013.

[DUR 93] DURAND G., *Les structures anthropologiques de l'imaginaire*, Dunod, Paris, 1993.

[DUR 00] DURAND G., *Introduction à la mythodologie*, Librairie générale française, Paris, 2000.

[DUR 15] DURKHEIM E., *Représentations individuelles et représentations collectives*, FB Éditions, Paris, 2015.

[EAS 08] EASTON L., SCHROEDER R., *The Influence of Imagination: Essays on Science Fiction and Fantasy as Agents of Social Change*, McFarland & Company, Jefferson, 2008.

[EIZ 74] EIZYKMAN B., *Science-fiction et capitalisme, critique de la position de désir de la science*, Mame, Tours, 1974.

[ELL 90] ELLUL J., *La technique ou l'enjeu du siècle*, Economica, Paris, 1990.

[ELL 04] ELLUL J., *Le système technicien*, Le Cherche-Midi, Paris, 2004.

[ENR 03] ENRIQUEZ E., *L'organisation en analyse*, PUF, Paris, 2003.

[FLI 95] FLICHY P., *L'innovation technique: Récents développements en sciences sociales, vers une nouvelle théorie de l'innovation*, La Découverte, Paris, 1995.

[FLI 01] FLICHY P., *L'imaginaire d'Internet*, La Découverte, Paris, 2001.

[FRE 00] FREEDMAN C., *Critical Theory and Science Fiction*, Wesleyan University Press, Middletown, Connecticut, 2000.

[GAR 12] GARÇON A.F., *L'imaginaire et la pensée technique: Une approche historique, XVIe-XXe siècle*, Éditions Classiques Garnier, Paris, 2012.

[GAU 87] GAUDIN T., *Le temps du germe, philosophie de l'innovation*, Strasbourg, Sophon, 1987.

[GEO 00] GEOFF K., *Science Fiction Cinema: From Outerspace to Cyberspace*, Wallflower, London, 2000.

[GOD 94] GODET M., *From Anticipation to Action: A Handbook of Strategic Prospective*, UNESCO Publishing, Paris, 1994.

[GOI 02] GOIMARD J., *Univers sans limites, Critique de la science-fiction*, Pocket, Paris, 2002.

[GOL 95] GOLDEN K.L., *Science Fiction, Myth, and Jungian Psychology*, The E. Mellen Press, Lewiston, 1995.

[GON 13] GONZATTO R.F., "The ideology of the future in design fictions", *Digital Creativity*, vol. 24, no. 1, pp. 36–45, 2013.

[GRA 13] GRAS A., *Les imaginaires de l'innovation technique*, Manucius, Paris, 2013.

[GRE 15] GREGORY A.P.R., *Science Fiction Theology: Beauty and the Transformation of the Sublime*, Baylor University Press, Waco, 2015.

[GRE 94] GRENIER C., *La science-fiction, lectures d'avenirs?*, Presses Universitaires de Nancy, Nancy, 1994.

[GRE 01] GRESCH L.H., WEINBERG R.E., *The Computers of Star Trek*, BasicBooks, New York, 2001.

[GRE 02] GRESH L.H., *The Science of Superheroes*, J. Wiley & Sons, Hoboken, NJ, 2002.

[GUN 05] GUNN J.E., CANDELARIA M., *Speculations on Speculation: Theories of Science Fiction*, Scarecrow Press, Lanham, 2005.

[HAL 13] HALES D., "Design fictions an introduction and provisional taxonomy", *Digital Creativity*, vol. 24, no. 1, pp. 1–10, 2013.

[HAN 06] HANEY W.S., *Cyberculture, Cyborgs and Science Fiction: Consciousness and the Posthuman*, Rodopi, Amsterdam, 2006.

[HAT 08] HATCHUEL A., WEIL B., LE MASSON P., *Les processus d'innovation: Conception innovante et croissance des entreprises*, Hermès-Lavoisier, Paris, 2008.

[HAY 99] HAYLES N.K., *How We Became Posthuman: Virtual Bodies in Cybernetics, Literature, and Informatics*, University of Chicago Press, Chicago, IL, 1999.

[HEU 03] HEUSER S., *Virtual Geographies: Cyberpunk at the Intersection of the Postmodern and Science Fiction*, Rodopi, Amsterdam, 2003.

[HIG 04] HIGGINS M., LIGHTFOOT G., PARKER M. *et al.*, *Science Fiction and Organization*, Taylor & Francis, Abingdon-on-Thames, 2004.

[HIP 88] HIPPEL E., *The Sources of Innovation*, Oxford University Press, New York, 1988.

[HOL 02] HOLLINGER V., GORDON J., *Edging into the Future: Science Fiction and Contemporary Cultural Transformation*, University of Pennsylvania Press, Philadelphia, 2002.

[HOT 00] HOTTOIS G., *Philosophie et science-fiction*, Vrin, Paris, 2000.

[HOT 13] HOTTOIS G., *Généalogies philosophique, politique et imaginaire de la technoscience*, Vrin, Paris, 2013.

[HUB 88] HUBKA V., EDER W.E., *Theory of Technical Systems. A Total Concept Theory for Engineering Design*, Springer, Berlin, 1988.

[ICH 05] ICHBIAH D., *Robots: From Science Fiction to Technological Revolution*, Harry N. Abrams, New York, 2005.

[JAM 91] JAMESON F., *Postmodernism or the Cultural Logic of Late Capitalism*, Verso, London, 1991.

[JOH 11] JOHNSON B.D., *Science Fiction for Prototyping: Designing the Future with Science Fiction*, Morgan & Claypool Publishers, San Rafael, 2011.

[JUN 90] JUNG C.G., *The Archetypes and the Collective Unconscious*, Princeton University Press, Princeton, 1990.

[KER 07] KERSLAKE P., *Science Fiction and Empire*, Liverpool University Press, Liverpool, 2007.

[KIN 00] KING G., KRZYWINSKA T., *Science Fiction Cinema: From Outerspace to Cyberspace*, Wallflower Press, London, 2000.

[KIR 10] KIRBY D., "The future is now: diegetic prototypes and the role of popular films in generating real-world technological development", *Social Studies of Science*, vol. 40, no. 1, pp. 41–70, 2010.

[KRE 86] KREUZIGER F.A., *The Religion of Science Fiction*, Bowling Green State University Popular Press, Bowling Green, OH, 1986.

[KUH 79] KUHN T.S., *The Structure of Scientific Revolution*, University of Chicago Press, Chicago, 1979.

[LAB 97] LABOUNOUX G., *Malaise dans l'organisation: le pouvoir imaginaire*, L'Harmattan, Paris, 1997.

[LAF 04] LAFONTAINE C., *L'empire cybernétique*, Le Seuil, Paris, 2004.

[LAN 06] LANGLET I., *La science-fiction: Lecture et poétique d'un genre littéraire*, Armand Colin, Paris, 2006.

[LAT 04] LATOUCHE S., *La mégamachine: raison technoscientifique, raison économique et mythe du progrès: essais à la mémoire de Jacques Ellul*, La Découverte, Paris, 2004.

[LAT 89] LATOUR B., *La science en action*, La Découverte, Paris, 1989.

[LEG 06] LEGROS P., MONNEYRON F., RENARD J.-B. *et al.*, *Sociologie de l'imaginaire*, Armand Colin, Paris, 2006.

[LET 08] LE THEULE M.-A., FRONDA Y., "La créativité dans l'entreprise", in UZUNIDIS D. (ed), *L'innovation et l'économie contemporaine. Espaces cognitifs et territoriaux*, Editions De Boeck Université, Brussels, pp. 147–194, 2008.

[LUO 13] LUOKKALA B.B., *Exploring Science through Science Fiction*, Springer Science & Business Media, Berlin, 2013.

[MAL 91] MALMGREN C.D., *Worlds Apart: Narratology of Science Fiction*, Indiana University Press, 1991.

[MAL 00] MALRIEU P., *La construction des imaginaires*, L'Harmattan, Paris, 2000.

[MAN 06] MANNHEIM K., *Idéologie et utopie*, Ed. de la Maison des sciences de l'homme, Paris, 2006.

[MAR 95] MARCUS G.E., *Technoscientific Imaginaries*, University of Chicago Press, Chicago, 1995.

[MAR 13] MARKUSSEN T., KNUTZ E., "The poetics of design fiction", *Proceedings of the 6th International Conference on Designing Pleasurable Products and Interfaces*, DPPI'13, p. 231, 2013.

[MAS 16] MASURE A., "Panne des imaginaires technologiques ou design dans un monde réel?", in DUNYACH I., PANDELAKIS P. (eds), *Actes de la journée d'étude CinéDesign: pour une convergence disciplinaire du cinéma et du design*, Université de Toulouse, 2016.

[MAT 00] MATTELART A., *Histoire de l'utopie planétaire. De la cité prophétique à la société globale*, La Découverte, Paris, 2000.

[MCC 97] MCCURDY H., *Space and the American Imagination*, Smithsonian Institution Press, Washington, 1997.

[MIC 08] MICHAUD T., *Télécommunications et science-fiction*, Editions Memoriae, Paris, 2008.

[MIC 10a] MICHAUD T., *La stratégie comme discours, la science-fiction dans les centres de recherche et développement*, L'Harmattan, Paris, 2010.

[MIC 10b] MICHAUD T., *Prospective et science-fiction*, L'Harmattan, Paris, 2010.

[MIC 14a] MICHAUD T., *L'imaginaire et l'organisation: la stimulation de l'innovation technoscientifique par la science-fiction*, L'Harmattan, Paris, 2014.

[MIC 14b] MICHAUD T., "La dimension imaginaire de l'innovation: l'influence de la science-fiction dans la construction du cyberespace", *Innovations, Revue d'économie et de management de l'innovation*, vol. 44, pp. 213–233, 2014.

[MIC 17] MICHAUD T., *Innovation, Between Science and Science Fiction*, ISTE Ltd, London, and Wiley, New York, 2017.

[MIC 18] MICHAUD T., *La réalité virtuelle, de la science-fiction à l'innovation*, L'Harmattan, Paris, 2018.

[MIN 96] MINOIS G., *Histoire de l'avenir, des prophètes à la prospective*, Fayard, Paris, 1996.

[MIN 16] MINVIELLE N., WATHELET O., MASSON A., *Jouer avec les futurs: utilisez le design fiction pour faire pivoter votre entreprise*, Pearson Education, Paris, 2016.

[MOY 00] MOYLAN T., *Scraps of the Untainted Sky: Science Fiction, Utopia, Dystopia*, Westview Press, 2000.

[MOL 90] MOLES A., "La fonction des mythes dynamiques dans la construction de l'imaginaire social", *Cahiers de l'imaginaire*, vol. 5/6, pp. 9–33, Boulder, 1990.

[NEL 82] NELSON R.R., WINTER S.G., *An Evolutionary Theory of Economic Change*, Harvard University Press, Cambridge, MA, 1982.

[NOB 97] NOBLE D.F., *The Religion of Technology, the Divinity of Man and the Spirit of Invention*, A. A. Knopf: Distributed by Random House, New York, 1997.

[NOV 14] NOVA N., *Futurs? La panne des imaginaires technologiques*, Les moutons électriques, Montélimar, 2014.

[OSB 53] OSBORN A.F., *Applied Imagination: Principles and Procedures of Creative Thinking*, Charles Scribner's Sons, New York, 1953.

[PAR 96] PARKER C., *Sci-fi and the Internet: Future Visions*, Future, Avon, 1996.

[PAR 92] PARKER M. (ed.), *Utopia and Organization*, Blackwell, Oxford, 1992.

[PAR 01] PARRINDER P., *Learning from Other Worlds: Estrangement, Cognition and the Politics of Science Fiction and Utopia*, Duke University Press, Durham, 2001.

[PIN 03] PINSKY M., *Future Present: Ethics and/as Science Fiction*, Fairleigh Dickinson University Press, Madison, NJ/Associated University Press, London, 2003.

[RIC 05] RICOEUR P., *L'idéologie et l'utopie*, Le Seuil, Paris, 2005.

[SAL 07] SALMON C., *Storytelling: La machine à fabriquer les histories et à formater les esprits*, La Découverte, Paris, 2007.

[SCH 89] SCHUMPETER J.A., *Business Cycles: A Theoretical, Historical and Statistical Analysis of the Capitalist Process*, Porcupine Press, Philadelphia, PA, 1989.

[SCH 06] SCHUMPETER J.A., *Capitalisme, socialisme et démocratie*, Payot, Paris, 2006.

[SCH 94] SCHOLES R.E., *Structural Fabulation: An Essay on Fiction of the Future*, UMI, Books on demand, Ann Arbor, MI, 1994.

[SEG 05] SEGAL H.P., *Technological Utopianism in American Culture*, Syracuse University Press, New York, 2005.

[SEG 92] SEGRESTIN D., *Sociologie de l'entreprise*, Armand Colin, Paris, 1992.

[SER 15] SERVIGNE P., STEVENS R., *Comment tout peut s'effondrer. Petit manuel de collapsologie à l'usage des génération présentes*, Le Seuil, Paris, 2015.

[SFE 95] SFEZ L., *La santé parfaite, critique d'une nouvelle utopie*, Le Seuil, Paris, 1995.

[SFE 02] SFEZ L., *Technique et idéologie, un enjeu de pouvoir*, Le Seuil, Paris, 2002.

[SHE 12] SHEDROFF N., NOESSEL C., *Make It So: Interaction Design Lessons from Science Fiction*, Rosenfeld Media, New York, 2012.

[STA 06] STABLEFORD B., *Science Fiction and Science Fact, an Encyclopédia*, Routledge, New York, 2006.

[STA 07] STABLEFORD B., *Sociology of Science Fiction*, Borgo Press, San Bernardino, 2007.

[STO 00] STOCKWELL P., *The Poetics of Science Fiction*, Longman, London, 2000.

[SUV 79] SUVIN D., *Metamorphosis of Science Fiction: On the Poetics and History of a Literary Genre*, Yale University Press, New Haven, 1979.

[WAR 82] WARRICK P.S., *The Cybernetic Imagination in Science Fiction*, MIT Press, Cambridge, 1982.

[WEB 98] WEBER M., *L'éthique protestante et l'esprit du capitalisme*, Pocket, Paris, 1998.

[WEI 79] WEICK K.E., *The Social Psychology of Organizing*, McGraw-Hill, New York, 1979.

[WEI 95] WEICK K.E., *Sensemaking in Organizations*, Sage Publications, Californie, Thousand Oaks, 1995.

[WES 98] WESTFAHL G., *The Mechanics of Wonder: The Creation of the Idea of Science Fiction*, Liverpool University Press, Liverpool, 1998.

[WES 00] WESTFAHL G., *Space and Beyond: The Frontier Theme in Science Fiction*, Greenwood Publishing Group, Santa Barbara, 2000.

[WOO 02] WOOD A., *Technoscience in Contemporary American Film: Beyond Science Fiction*, Manchester University Press, Manchester, 2002.

8

Science Fiction, Innovation and Organization: Where Do We Stand?

8.1. Introduction

The literature on innovation is very large and it includes many subfields. Researchers started exploring innovation as the process of the implementation of invention roughly at the end of the Second World War. Rogers' [ROG 62] seminal work on both the process and the actors of innovation paved the way to a lot of work on various aspects of the adoption, diffusion and to a lesser extent, rejection of innovation [ABR 91, CUS 02, KNI 67, MUR 07, MYE 69, ROT 94, TEE 86, VON 86]. Dodgson *et al.* [DOG 05] produced a typology of the range of research in innovation. They identified six major focuses of innovation research: the source of innovation, nature and extent, type, system, process and finally the outcomes. For a long time, research has sought to identify the keys to successful innovation and the factors that could encourage their rejection or failure [QUI 85, SAN 77, VOS 85]. These studies shared the conclusions that the market, as the technological advance as the structure of organizational processes and management practices are critical elements [VAN 03]. This was probably the reason why research has historically focused on innovation in how to manage innovation processes, and more broadly in organizational and managerial innovations [VOL 13]. But, while the management of technological and service innovation has long been studied, management innovation has received much more recent and moderate attention, as shown by two bibliometric studies [CRO 10,

Chapter written by Sonia ADAM-LEDUNOIS, Claire AUPLAT and Sébastien DAMART.

KEU 12]. Today, research in innovation has a wide scope including that of technologies, services, organizations and management.

Science fiction has historically been built on the basis of a fascination with science, technological progress and innovation. The science fiction subgenre known as hard science fiction (or hard SF) is the heir to this historical foundation. Other categories of science fiction works are based on a rejection of technology, which inspires fears and fantasies. Dystopia is another subcategory of the science fiction genre. It has gradually developed according to a rejection of social organizations that were contemporary to them, ranging from the hardest capitalism to radical bureaucratic and socialist planning. This is the subject of the contribution presented in this chapter. The spectrum of science fiction works is very broad and it is allowed to take stock of what science fiction has brought both in its dimension of fascination and projective as in that of criticism and rejection to the vision of the processes of innovation and the emergence of new forms of organization. This state of the art is the subject of the contribution presented in this chapter.

To build this state of the art, we need to make a return on the diversity of forms taken by science fiction works. In doing so, we can understand the analyses by recent decades research on science fiction and its contributions to the critical analysis of technological and organizational innovation processes. The remainder of the chapter is then structured as follows: the next section situates and defines science fiction. Then we explain the methodology we used to build the synthesis of academic works presented in this chapter. The following section presents the major findings of the literature review, and the last part is devoted to discussion and conclusions.

8.2. Science fiction in its diversity

Science fiction belongs to popular culture. Multiple types of protagonists have deliberately or not participated in the work of structuring and legitimizing the genre (readers, authors, critics, academics, scientists, fans, communities developed around series, films, etc.). The field is in a perpetual state of incompletion [EVA 99]. There are institutional reasons for this, e.g. the place of science fiction in teaching, in academy, in literary journals or in the media, but there are also reasons related to the extreme heterogeneity of science fiction works both in the literature (novels, comic strips, etc.) and in the cinema (films, series, cartoons, etc.). Marg Gilks and Moira Allen completed in 2003 a census of science fiction subgenres: apocalyptic, holocaust, post-apocalyptic, cyberpunk, first contact, hard science fiction, military science fiction, near future science fiction, fantasy science, slipstream, soft/sociological science fiction, space opera and time travel [GIL 03]. Browsing the Websites of science fiction communities or different fanzines easily convinces us that this list is not exhaustive and can also include afrofuturism, biopunk, feminist science fiction, libertarian science fiction, mathematical science fiction, mundane science fiction, space western, etc.

The recording of the origins of science fiction seems to be a matter of personal opinion, and some researchers date it back to utopian works written between the 15th and 18th Centuries [ROB 16b]. However, some consensus emerges around the pioneering dimension of works by Jules Vernes, H.G. Wells, Edgar A. Poe or Mary Shelley. Likewise, even if it is complicated to find consensus around major works of science fiction, some references are usually considered unavoidable. Let us cite for the cinema: *Metropolis* (1927, Fritz Lang), *Forbidden Planet* (1956, Fred M. Wilcox), *Planet of the Apes* (1968, Franklin J. Schaffner), *2001: a Space Odyssey* (1968, Stanley Kubrick), *Star Wars* (1977, Georges Lucas), *Alien* (1979, Ridley Scott), *Mad Max* (1979, Georges Miller), *Blade Runner* (1982, Ridley Scott), *Terminator* (1984, James Cameron), *Back to the Future* (1985, Robert Zemeckis), *Total Recall* (1990, Paul Verhoeven), *The Matrix* (1999, Lana and Andy Wachowski), *Inception* (2010, Christopher Nolan) and *Interstellar* (Christopher Nolan, 2014). The list is in itself interesting as these films all share a focus on the future and a form of mix between wonder and pessimism, which are characteristic of contemporary science fiction [PAN 89, KLE 77].

Researchers, literary historians and science fiction authors themselves have tried many times to define the genre of science fiction and this has led to the definition of science fiction as a field of controversy [ROB 16a]. Most authors agree that science fiction is a very heterogeneous set of works relying mostly on complex and endless combinations of a different form of society with alien creatures in various forms [KIN 03]. This heterogeneity is a difficulty and has given rise to many constructions of criteria to delimit the genre.

Hugo Gernsback, the inventor of the term "science fiction" (scientifiction), proposed in a 1926 editorial of his brand new magazine *Amazing Stories* to consider that science fiction was characterized by three fundamental common components – narrative, scientific information and prediction; three audiences – general readers, young people and scientists; and three types of objectives – to entertain readers, to provide scientific education and to offer stimulating ideas to scientists and inventors. The predominance of science in science fiction probably distinguishes works published before the Gernsback era from those written and published later [WES 92].

Russ [RUS 75] noted that the plausibility of fiction was a trait that characterized science fiction in the first classifications proposed by academics like Darko Suvin – a critic and academic from Croatia and Quebec – or by authors and critics like Stanislas Lem – the Polish author of Solaris in 1961 – and Samuel Delany – an American author and academic. According to these typologies, science fiction should not be confused with fantasy works, even if the intersection with these works cannot be empty given the complex connections between science fiction and utopia and dystopia [WIL 78]. The notion of plausibility in science fiction is complex. More than the use of real-world references, it means the author uses text codes that allow them to build a coherent framework. When the frame is set, the reader can use his/her own experience in the real world to infer assumptions about how the fictional world works [EMM 97, STO 00].

In fact, defining science fiction is a task whose complexity arises in part from the fact that there are several paradigms to consider when dealing with the question of the literary genre in general. Thus, to define science fiction, we propose to refer to two types of approaches. One is comprehensive and consists of establishing a list of characteristics of what science fiction work is. The other anchors the work of science fiction in a social context, with specific rhetoric (and therefore symbolism) and values.

There are very few attempts to draw lists of what characterizes science fiction as the task seems so risky. Csicsery-Ronay [CSI 96] proposed seven hypotheses about what makes a piece of work science fiction. Even though the author himself admits that some of these hypotheses are questionable and that they do not form a complete and exhaustive system, they provide the basis for a repository and help us delimit an acceptable perimeter of the territory of science fiction. (1) Neologism: science fiction invents words and refers to new realities; (2) novums: science fiction relies on inventions or discoveries that have changed the course of history; (3) historical extrapolation: science fiction provides an explanation of the path that leads from our present to a certain future; (4) oxymoron: at the heart of the tale, there is an "absurd logical contradiction"; (5) scientific impertinence: the work of science fiction partly violates scientific laws, not to criticize the state of knowledge but to construct a situation with intriguing and dramatic dimensions; (6) sublime chronotopes: science fiction creates space-time that works with its own laws; (7) parable: science fiction works are vehicles of a certain morality, which may have a link with science and technology.

Within the approach that sees the genre as an historical process, Rieder [RIE 10] makes five propositions. First, science fiction is both historically situated and evolving. In this kind of approach, science fiction works are understood only if the context in which they were produced is documented. In other words, science fiction in this perspective is not just a literary technique. It is a social construct anchored in a social context. This approach is the one adopted in a large number of works in management that have used the study of science fiction works to understand concepts of organization and management that are rooted in a complex historical reality [HIG 01]. For example, studies about Gene Roddenberry's *Star Trek* TV shows and movies teach a lot about change in leadership and organizational discourse since the early 1970s [KAV 01]. Science fiction is thus anchored in a cognition of science and techniques at a given moment. Second, there are no unifying features, but rather a network of similarities and relationships between science fiction works. This point makes it difficult if not impossible to identify the point of origin of science fiction. Third, science fiction is not constructed by definitions that can be given of it, but rather by the uses made of its works (thus including authors, readers, science fiction communities, fans, etc.). Fourth, the identity of science fiction is constructed by positioning it in an "economy of genres". There, Rieder draws from Todorov [TOD 90] for whom genre is the institutionalized codification in a given society of some discursive properties. He considers the "economy of genres" as the fact for a reader to interpret generic codes of a text according to values and a reception strategy. The economy of genres obviously depends on all the genres at play but also on the values mobilized by the reader. Fifth, the attribution of the science fiction genre is a labeling and rhetorical activity. Thus, if a piece of work is labeled as science fiction, it is because it is understood that it must be read or seen using the codes of science fiction. However, the question of values in science fiction goes further than just a question of labels. In an interview he gave in 2018, Kim Stanley Robinson (author of *Red Mars*, and more recently of *New York 2140*) gave science fiction a simple and rich definition: "This is simplistic, but science is where we establish facts and fiction is where we establish values. The name 'science fiction' is very powerful because it seems to say we can bridge the fact/value conundrum" [STA 18, p .88]. The definition suggests that science fiction connects facts (science) with value systems. It puts the present as it is and a future that could exist according to a system of values. Kim Stanley Robinson's *Red Mars* (1992) is iconic. The novel provides a detailed description of the future of the conquest of Mars given a state of our relationship to science, technology and the capitalist system. The conflict between the first hundred – the first settlers on Mars, essentially a contingent of scientists – and transnationals can be seen as a consequence of unbridled and ethically fragile scientific progress but also of a borderless capitalism whose power is much greater than that of states.

8.3. A focused review of academic literature on science fiction: method

Our methodology is based on Tranfield *et al.*'s [TRA 03] Systematic Literature Review (SLR) method. As explained in [DEN 04], literature reviews in management studies were often traditionally completed in a descriptive and narrative manner, which weakened their generic value. Indeed, because of the ever-growing amount of published work on any given topic, these literature reviews might contain biases that left room for contradicting results from other research. Building on an approach that originated in the United Kingdom among the medical profession [MUL 94, THO 05b], SLRs are intended to minimize the risk of contradictory findings by embedding clear specifics of transparency and ideation pathways. The most important characteristic of an SLR is that "the [investigation] process is reported openly in the same way that empirical research would be" [PIT 04, p. 480]. In an SLR, the authors try to make their thought process clear and transparent by having a detailed methodology section in which they explain step by step which data they use and how they proceed to reach their interpretations and conclusions [THO 05b].

We first use keywords to filter data and constitute the corpus. We then analyze the results manually by reading abstracts or full papers to ground our discussion. We choose the same keywords and the same coding to make sure that we do not introduce biases in our research. To study academic papers dealing with science fiction, we choose a combination of two data bases: EBSCO Business Source and JSTOR. These two data bases are complementary as they cover fairly distinct fields of academic literature of humanities and social sciences. EBSCO Business Source, as indicated in the name, is devoted to business sources and it included at the beginning of 2018 over 2,000 peer-reviewed journals in the fields of accounting, administration, banking, economics, finance, human resources, management and marketing (source: EBSCO Business Source website). JSTOR is a digital library that provides full-text searches of more than 2,600 top scholarly journals in the fields of humanities, social sciences and sciences. EBSCO Business Source is therefore more focused on Business administration while JSTOR is more general in the humanities and social sciences.

Scopus and Web of Science were two alternative data bases that also cover scientific journals. However, we rejected them because they only provide abstracts and citation indexes, and we wanted to have access to full texts.

The first search keyword we used was science fiction, but we used both "science fiction" and "science fiction" terminologies to make sure we did not "miss" results. However, this brought a problem as in some searches "science fiction" was considered by search engines as a request to find "science" and "fiction" and produced irrelevant results. For these search engines, it meant that the two words appeared in the same book or journal, but did not automatically refer to science fiction. We had to manually eliminate the results that were inappropriate, i.e. where the proximity of the words science and fiction did not mean that the book or journal dealt with science fiction. We also had to eliminate duplicates from the databases. Finally, we cross-checked our findings with ScienceDirect – a leading platform of peer-reviewed scholarly literature that also contains open access sources – in order to verify that we had "caught" perhaps not all, but most of the papers we were looking for. This manual cross-checking allowed us to identify only a few more papers and so gave us confidence that we had "caught" the large majority of relevant papers in our systematic review.

In coherence with our objective, our list of search keywords was made with science fiction, innovation, management and organization. One of our methodological questionings concerned "technology". Should we add a "technology" keyword to our search keywords? We opted against it because the word "innovation" is naturally connected to "technology" and we felt it would introduce a bias that might bring distortion to our analysis as we considered a broad view of innovation alternatively applied to technology, services, products, processes, organization or management.

We checked whether the search engines made distinctions between organisation (with an s) and organization (with a z) on the one hand, and between science fiction (without hyphen) and science-fiction (with hyphen). Although this was time consuming, we felt it was important to carry out these checks. We did not find any difference with the two spellings of organization, and we did "catch" a few (<10) separate occurrences by running our search first with science-fiction (with the hyphen) and then science fiction (without the hyphen). We therefore concluded that the differences in spelling did not bring any significant results.

The EBSCO database produced 269 hits on the general search we carried out using the filters "English language", "academic publications", "science fiction" or "science-fiction". We proceeded slightly differently with the JSTOR database by using the same filters "English language", "academic publications", "science fiction" or "science-fiction" but only in titles while in EBSCO the search was in the main text of the papers as well. Our reason for doing so was that when we did not restrict our search to titles, we had a very large number of irrelevant returns that prevented us from reaching any valuable conclusion. We considered this restriction did not really alter the quality of our research because the ensemble of papers containing the word "science fiction" or "science-fiction" in their title was included in the larger ensemble of academic papers containing the term "science fiction" or "science-fiction" in the body of the text. On the other hand, the search done in the EBSCO database using the filters "science fiction" or "science-fiction" in titles only produced 53 returns, which we considered too few to be of interest. Restricting the filters to "titles only", we obtained a database of 565 journal articles in JSTOR. In order to allow a relevant analysis of the articles resulting from the search in the EBSCO, JSTOR and ScienceDirect databases, we proceeded in the following way. Potentially redundant articles between the three databases were identified and duplicates were removed. Then a manual search based on the reading of abstracts and of most of the contents led us to discard some articles that had no link with "organization", "management" or "innovation". Other articles did not deal with science fiction at all but contained the term "science fiction" used as a qualifying adjective. These items were also removed. Finally, we had a corpus of 105 articles.

The reading of these papers led us to adopt two classification dimensions that we organized as a matrix to try and elicit the potential markers of science fiction as an institutional agent. One dimension separated articles in which science fiction was analyzed as a vector of change ("transforming") from articles in which science fiction was considered as a description of a reality ("describing"). The term "reality" includes virtual reality since we are in the world of science fiction. Naturally, any work of fiction has a power of influence over a system. We therefore distinguished the articles according to whether this transformation capacity was explicitly studied by the authors of the article or not. The second dimension of our classification made it possible to sort the articles according to the object on which the capacity of transformation/description of science fiction as envisaged in the article bore. We identified four possible objects: theory, behaviors, organizations and technology/products/services.

When papers were assigned to the "theories" category, there were two possible reasons: (1) the papers analyzed how science fiction was used to assess existing theories on organization, on innovation management, etc., or to question traditional representations of organization and management; (2) the papers analyzed how science fiction was used to generate new possible explanations of reality, new theories.

In papers assigned to the "behaviors" category, we placed either: (1) studies where science fiction considered existing behaviors as possibly explained by former science fiction works (or where science fiction helped to assess the effects of current behaviors); or (2) studies that used science fiction in order to help anticipate what would become common behaviors (for example future behaviors of consumers).

When papers were placed in the "organizations" category, this meant that they were either: (1) studies in which science fiction assessed existing forms of organizations (or papers in which authors tried to assess science fiction in its ability to anticipate our current forms of organization); or (2) studies of the use of science fiction as a designing tool generating new forms of organizations. In both cases, the word "organization" was taken in its broad sense, including all forms of social entities.

In the "technology/product/service" category, we placed papers in which authors studied either: (1) the ability of science fiction to anticipate our current technologies and services (did *Star Trek* predict our use of smartphones nowadays?); or (2) science fiction as a prototyping tool that helps to design new technologies, new products or new services (including new business models).

8.4. Systematic literature review: findings

As shown in Figure 8.1, business publications (EBSCO) picked up on the science fiction topic with a gap of nearly 30 years compared to the more generalist human and social sciences, but since the early 2000s, both scientific domains devoted as much importance to it.

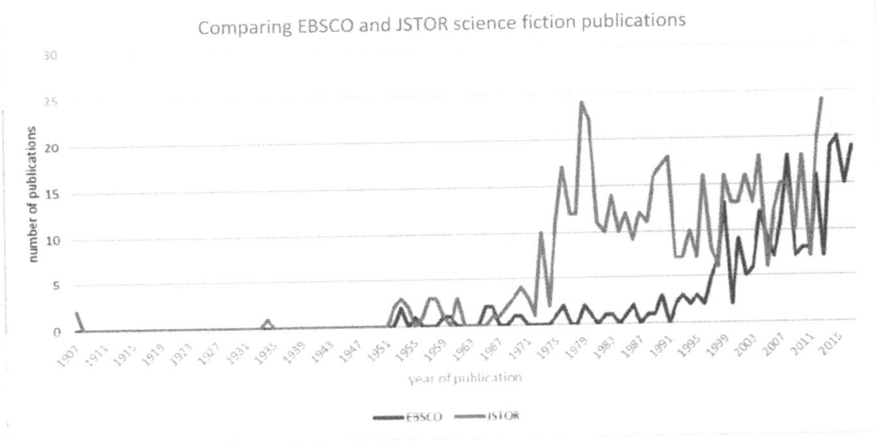

Figure 8.1. *Evolution of academic publication on science fiction. For a color version of the figure, see www.iste.co.uk/michaud/sciencefiction.zip*

In the EBSCO database, the search with the keywords: "science fiction and innovation", "science fiction and management", "science fiction and organization"; "science-fiction and innovation", "science-fiction and management", "science-fiction and organization" produced a corpus of 88 papers once duplicates had been removed, and the same search in the JSTOR database produced a corpus of 127 papers without duplicates. Once again, as shown in Figure 8.2 we noted a gap of several decades between the start of publications in JSTOR and in EBSCO. Besides, the interest for the topics was more constant in JSTOR publications. In EBSCO, it is oscillating between high peaks and periods of lesser interest.

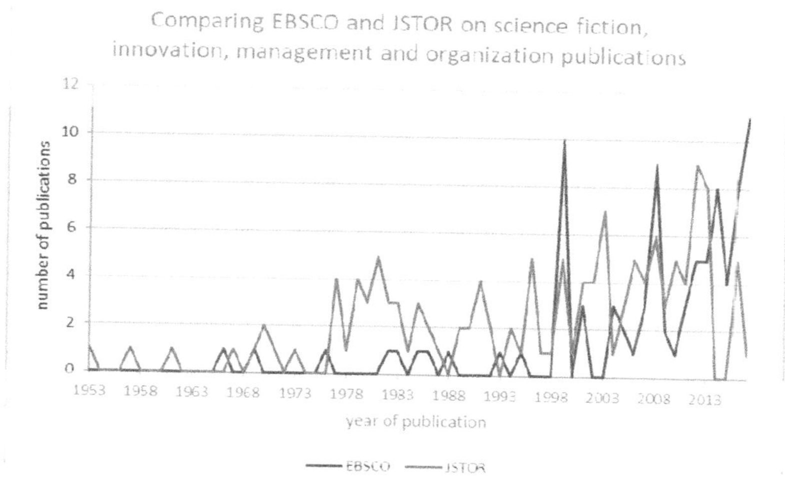

Figure 8.2. *Evolution of academic publication on science fiction filtered on innovation, management and organization. For a color version of the figure, see www.iste.co.uk/michaud/sciencefiction.zip*

The academic literature dealing with the science fiction genre is beginning in the early 1950s. Meanwhile, the genre is entering its golden age. Books flourish, but the field of research on these works, what they tell, how they are constructed, what they say about contemporary societies is still in a structuring phase. The journal *Science Fiction Studies* will only be designed in 1973 by R.D. Mullen. It is also from the beginning of the 1970s that we list the first academic articles questioning the relationship between science fiction, technology and organization.

As shown in Table 8.1, the papers from our systematic review belong overwhelmingly to the "describing" category. They focus on science fiction as a tool to evaluate a virtual world and they analyze the characteristics of these virtual worlds. The most represented categories (highlighted in gray) correspond to the articles whose study concerns science fiction as a tool to describe either existing technologies/products/services or existing organizational forms. Science fiction is analyzed much less frequently as a tool for the direct and explicit transformation of a reality.

The results of assigning articles to one of the eight categories (2 × 4) are shown in Table 8.1.

	Theories	Behaviors	Organizations	Tech/products/ services	Total
Describing	14	4	23	27	68
Transforming	7	2	7	21	37
Total	21	6	30	48	105

Table 8.1. *Sorting of the 105 papers (results)*

	Theories	Behaviors	Organizations	Products/services
Describing	[ANG 79, ATT 05, BLO 99, CAS 99, DON 08, EDW 14, ESP 17, KEA 99, NIC 04, PAK 16, PHI 99, REN 80, SEE 10, SUV 82]	[JOH 09, MAR 13, SUV 81, TUR 14]	[ABB 12, BAR 12, BEL 95, BLA 86, BYE 87, CSI 03, CZA 08, DOH 89, FRA 83, GOV 84, HAM 12, HIG 12, HIG 13, KEN 69, LEF 12, LEM 85, MA 13, NUD 89, PAR 08, REE 08, RIV 12, THO 05a, ZIS 13]	[BAN 03, BER 05, BER 79, BRO 07, COB 05, DE 13, DON 04, FLA 99, FOS 16, FRE 98, GRA 94, HAM 03, LAU 11, NIU 08, RAB 11, SCH 17, SCH 18, SCH 99, SII 96, SON 13, STE 14, TEL 98, VIN 17, WEB 08, WEB 16, YEO 14]
Transforming	[DE 09, DIN 15, GEN 17, PAR 99, SAM 96, STI 91, TIR 99]	[POT 14, STO 73]	[BIE 12, FEK 79, HY 76, PEP 17, SRI 99, WIL 78, WIT 08]	[ACK 15, ALE 09, ARC 17, BAT 17, BEL 13, BUR 15, COL 99, DEN 01, ESP 12, FEI 07, FLE 16, GOL 01, GRA 15, IDI 00, KUR 08, KYM 16, LOP 04, MIL 10, MOL 69, RAV 17, STE 17]

Table 8.2. *Academic contributions on science fiction filtered based on innovation, management and organization*

8.5. How science fiction sees technology and organizations

As shown in Table 8.1, the academic literature is more concerned by the links between science fiction and technology (innovative products and services) than with organization and management. Historically, science fiction was first built around technological innovation. It is not surprising that in our literature review, the topics covered by the articles focus more on technology: products, services and uses that are made today or will be made tomorrow. As previously mentioned, science fiction was born of an appetite for science and a belief in technological progress. Like Jules Verne's long, precise and documented descriptions of the Nautilus and its underwater environment in *20,000 Leagues Under the Sea*, published in 1869–1870, science fiction authors (and hard science fiction in particular) often develop a level of scientific and technological knowledge that is very high [SCH 18].

Organizations also play an important role in science fiction works. In many of them, organizations are present. They are not necessarily the central theme of the novel, film or science fiction series, but they often support the author's or director's demonstration. The literature shows that futures built by science fiction are translated by science fiction into different possible organizational and institutional forms (often dystopian). The literature shows that science fiction is a trace of the author's contemporary (and dominant) ideologies and (organizational) representations. As for technology, the literature finally shows that science fiction has a generative capacity for scenarios.

8.6. Dystopian visions of technologies and organizations

As used and discussed in science fiction works, technology can be defined as a "rational, systematic, taught, learned, and replicable way of materially controlling the material world, or parts of it" [RUS 78, p. 256]. The author adds that technology in science fiction has four basic characteristics. First, it is modern, a sign of progress made by man in the state of scientific knowledge. Second, it is ubiquitous. Technology is everywhere. Third, technology is autonomous. This term does not only refer to robots or artificial intelligence. Whatever it is, technology acts, influences, transforms and threatens. Finally, technology is not controllable and it controls the human being. In this, technology appears dangerous. It contains in itself some form of fatalistic damnation and a form of hope as well. There is a certain paradox there. Matheson [MAT 92] indicates that science fiction often shows technology as the possible instrument of human dominance when it is misused. However, science fiction also leaves the possibility to glimpse a certain form of human liberation because technology is the application of a systematic form of rationality to a complex reality.

Anyway, our literature review exposes that technology is very often used by science fiction in a dystopian perspective. Corbett [COR 95] shows that on a panel of 117 Anglo-American films released between 1970 and 1995, 39 of them are films in which the technology (its use or the technology itself) is directly responsible for a dystopian situation (or apocalypse). Post-apocalyptic works that postulate that man will use technology to destroy humanity are numerous [GRA 94]. We find this idea as a postulate of numerous science fiction novels through decades: Richard Matheson's *I Am Legend* (1954), Pat Franck's *Alas, Babylon* (1959), Robert Merle's *Malevil* (1972), Robert Silverberg's *Hot Sky at Midnight* (1994), Matthew Mather's *Cyber Storm* (2013), etc. It is also the cynical position adopted by character Senator Thurman in Hugh Howey's Silo trilogy written in 2013–2014. The senator is convinced that man will inevitably end up causing his own loss. Thurman anticipates this fall, provoking it with thermonuclear explosions. Before that, he organized the survival of a few hundred humans, in underground silos, kept for several centuries away from any form of technological progress and used nanotechnologies to prevent them from getting out of the silos. It is also technology that wrecked the human race in *Terminator 3* (2003, Jonathan Mostow) due to the takeover of the main defense centers by an artificial intelligence network (Skynet).

In extreme versions, technology is seen as having taken complete control of consciousness to the point of becoming invisible and thinking in place of human beings. In Peter Riva's *The Path* (2015), the "System", an AI massive program runs everything on earth. It includes thinking for the human being. The novel questions the existence of human beings (for whom the Latin proverb "*cogito ergo sum*" does not work anymore).

In the background of the relationship between nature and technology, science fiction exploits the fear that man will be replaced by technology. Scholars have very often explored the challenges of post humanism [GUT 14, SCH 17, SCH 99, STE 14]. Kurzweil [KUR 05] identified several scenarios. In the first one, humans are considered by machines too destructive and dangerous to justify their existence. In another scenario, technology is supposed to protect men from their weaknesses (and in particular, machines prevent men from yielding to their taste for self-destruction). In the most optimistic scenario, man has developed such a technological integration that he becomes a cyborg "by nature". This last scenario provokes a debate. Indeed, work around artificial intelligence seems to have had for decades the objective of reproducing human forms of reasoning. Kurzweil's third scenario suggests that humans can transcend their natural biological part for their own benefit. Science fiction plays on the three scenarios. The first two allow to build dramas and dystopias. The robots that take control of the city in the film version *I, Robot* (2004, Alex Proyas) of Harlan Ellison and Isaac Asimov's 1978 script, do so on the grounds that this takeover is in the interest of human beings. The Skynet network of *Terminator 3* is primarily a defense system. The third scenario of Kurzweil leads less easily to dystopia.

8.7. Highlighting ideologies behind technology and organizations

What are the ideologies and fundamental assumptions made in different sociotechnical systems? Science fiction provides answers to this question through innovation choices that carry visions of a more or less desirable future [SON 13]. The same applies for representations of social systems and production regimes [KEN 69, LEM 85]: an organization is built on the basis of choices made in terms of regulation of social relations.

Our literature review lists work that treats science fiction as a medium for criticizing typical organizational forms. Several studies demonstrate how science fiction captures systems' ideologies from the organizations that constitute or represent them. This may concern neoliberal capitalist systems [BYE 87, RIV 12, THO 05a] as well as totalitarian or postdictatorial systems dominated by bureaucratic and technocratic organizations on the model of the political organization described by George Orwell in *1984* [BEL 95, DOH 89, HAM 12, NUD 89]. As a literary genre, science fiction has often accurately described movements of resistance to the world of organizations, whether these organizations are inspired by a capitalist, a socialist or a totalitarian ideology. Srinivas [SRI 99] even takes up the resistance incarnated by the characters in Philip K. Dick's novels by applying it to managers who, in his opinion, must resist the movement of "androidization". Following this goal of denouncing the perverse effects of dominant ideologies, science fiction has taken a stand in the great value conflicts of the 20th Century. Many authors show that science fiction has regularly denounced various forms of imperialism – from states or from companies – [CSI 03] as well as movements of colonization and decolonization [HIG 13]. Kim Stanley Robinson's *Red Mars* (1992) is a careful deconstruction of colonization and decolonization movements. The ideological oppositions of large blocks are at the heart of many science fiction productions. James Corey's *The Expanse* (2014–2017) describes a solar system plagued by latent conflict between three major blocks: The Earth, Mars and the outer planets. In a way, it is a Cold War-like conflict that is implicitly the backdrop of many science fiction novels [BAR 12].

Our literature review shows that science fiction is an aggregator of scrutiny of power relationships between blocks or social groups and that it provides a fairly unified vision: technology is the means to acquire power supremacy.

In science fiction, the image of omnipotent technology is reinforced by a process of "monumentalization" of technology [TEL 98]. In cinema, as in literature, the process consists of a technique of "aestheticization" of machines and technology based on vanishing lines giving a particularly impressive perspective. Everything is big. The machines are of a size that far exceeds the human size (H.G. Wells' *The War of the Worlds*, 1898); the buildings are imposing and the cities are necessarily made up of skyscrapers (Isaac Asimov's *The Caves of Steel*, 1953); the roads no longer exist or they are in 3D (Philip K. Dick's *Do Androïds Dream of Electric Sheep?*, 1968). The process underscores the impossible control of technology by human beings because technology exceeds them in the proper sense as well as in the metaphorical sense. The process is related to the postcapitalist vision of firms, i.e. firms that develop and possess technologies. A parallel can be drawn between the representation of omnipresent and out of control technology and the company of the future also presented as omnipotent and out of reach. Numerous evocations of transnational (or trans planetary) groups have full powers beyond states: the U.S. Robotics and Mechanical Men Inc. of Isaac Asimov's novels, the transnationals of Kim Stanley Robinson's *Red Mars* (1992), etc. At the heart of these evocations of business and technology, it is power that is at stake and that drives innovation. Sometimes, technology is a means to assert power as in the image of the Capitol in Suzanne Collins' *Hunger Games* trilogy (2008–2010). Often technology is represented in an elitist version, owned by a minority, in a fragmented society, like in Richard Morgan's *Altered Carbon* (2002), a novel probably inspired by Robert Shekley's *Immortality Inc.* (1959), a science fiction thriller in which the richest have the possibility to implement their memory (a backup copy) in a completely new high-quality body (a sleeve). The poorest only have access to low-quality bodies or no body at all.

Organizations and technology are artifacts and as such they are in conflict with nature. This confrontation is an image of the permanent and ideological opposition of good and evil. Siivoen [SII 96] places the oppositions between good and bad, nature and culture, essentialism and culturalism, and nature and technology on the same level. The conflict between nature and technology is however more complex because technology is immersive [HUH 92]. Humans are immersed in technology because they live in a world of technology and are also subjected to this technology. Science fiction work on cyborgs and robots is exploring this paradox. Isaac Asimov's founding work on robots shows man's confusing and ambiguous relationship to cybernetics. Robots are enslaved (and the three fundamental laws of robotics ensure this in theory[1]) and at the same time, men use robots in their capacity to become more and more autonomous: to act alone, to think for themselves, to feel by themselves, to be endowed with consciousness.

Science fiction, through its explorations of the relationship between nature and technology, leads to metaphysical questions about what constitutes the deep nature of human beings. Robots, in a certain way, re-humanize man in that they lead him to reveal what man is specifically. Sims [SIM 09] shows the specificity of the use of technologies by humans: "Many animals also have, in various forms, the technology of language. A major difference between the human use of technology and that of other animals is that humans have an ongoing dialogue about what technology is; as a result, they can make modifications and sophistications on previously existing forms of technology within their own lifetime, which, though possible for other species, is unusual" (p. 68). This dialogue between technology and the human leads science fiction to constantly seek where the human is hiding behind what is not human. This is the case for works dealing with super-humans, anti-humans, in-humans, a-humans, etc., and this is the case of course with robots, androids, cyborgs, zombies, vampires, etc. [DOU 12]. It has little to do with the physical appearance of these non-human entities. Dougherty recalls that the artificial intelligence HAL in Stanley Kubrick's *2001: A Space Odyssey* has no face or anthropomorphic appearance but just

1 (1) "A robot may not injure a human being or, through inaction, allow a human being to come to harm.

(2) A robot must obey the orders given it by human beings except where such orders would conflict with the First Law.

(3) A robot must protect its own existence as long as such protection does not conflict with the First or Second Laws". [ASI 50, p. 40]

a camera lens and that despite this, the film director manages to treat the anxiety, mistrust, paranoia of the artifact HAL.

8.8. Science fiction as the source of new technological and organizational scenarios

Academic literature on science fiction and innovative technology illustrates a paradox. Despite the fact that technology is often seen by science fiction as the source of apocalypse, several studies show how science fiction can be used in a constructive design process. Science fiction is then considered as a prototyping tool [BEL 13, BUR 15, GRA 15, KYM 16]. Science fiction is also more simply a tool for detecting emerging technologies [ALE 09] and possibly a tool that intentionally or unintentionally influences citizens' perceptions of technological innovation. This is particularly manifest for technologies associated with robotics [DE 13] or even nanotechnologies [COB 05].

The academic literature that deals with science fiction and organizations tends to focus on the dystopian and negative nature of organizations' lives. The generative dimension of science fiction – its capacity to produce new forms of organization and new management practices – is not very present. Some articles, however, explore the generative capacity of science fiction, which is also applicable to the invention of new technologies. Some point out why science fiction can usefully contribute to the business world at the moment: not because it is predictive, but rather because it can help generate new frameworks of thought and because it questions fundamental assumptions of management and business [PEP 17]. Biello [BIE 12] shows that science fiction can provide new frameworks for thinking about energy resource management in the future. Witt and deHaven-Smith [WIT 08] engage in a similar exercise taking into account security issues in the post-9/11 world. However, this vision of science fiction as generative of new forms of organization concerns a small share of papers in the academic literature. Most often, authors write about science fiction that portrays companies as it does technology: dehumanized, out of control and monumental. We can also question the absence of current managerial innovations in science fiction work: how does science fiction treat the experiences of liberated companies like Zappos? What weight does science fiction give to the growing share of slashers and non-wage workers worldwide? A large number of science fiction works have in common to describe a relatively old version of organizations: based on classical hyperspecialization and the use of bureaucratic coordination methods.

8.9. Conclusion: three demonstrations and a possible research avenue

Works of science fiction have offered extremely varied representations of innovation and organizations. These are the two objects that have most inspired science fiction authors. Our systematic literature review of academic research on science fiction and its relationship to innovation, management and organizations provides three demonstrations. (1) Science fiction's attention to technology leads her to see it as an object that potentially exceeds the human. Science fiction has therefore helped to build an imaginary in which the human renounces some of its freedom, its margins and the social functioning does not prevent this. (2) Science fiction has positioned itself in revealing (and defending) ideologies masked behind different forms of organization by employing the same mechanism as that used for the technology: the monumentalization of the devices of organization (starting with the company, global and dominant) and the staging of their uncontrollable and out of reach nature. (3) Science fiction has a generative power for new technological and organizational scenarios but this power remains for the moment little mobilized. This third point draws the contours of a research program on the identification of science fiction's avenues of mobilization to generate new technological scenarios with positive impacts and to highlight new organizational and management levers.

8.10. References

[ABB 12] ABBOTT C., "Rocky mountain refuge: constructing 'colorado' in science fiction", *Science Fiction Studies*, vol. 39, no. 2, pp. 221–242, 2012.

[ABR 91] ABRAHAMSON E., "Managerial fads and fashions: the diffusion and refection of innovation", *Academy of Management Review*, vol. 16, no. 3, pp. 586–612, 1991.

[ACK 15] ACKERMAN E., "The race to build a real-life tricorder", *IEEE Spectrum*, vol. 52, no. 6, pp. 46–85, 2015.

[ALE 09] ALEXANDER B., "Apprehending the future: emerging technologies, from science fiction to campus reality, EDUCAUSE Review, 2009", *EDUCAUSE Review*, vol. 44, no. 3, pp. 12–29, 2009.

[ANG 79] ANGENOT M., SUVIN D., "Not only but also: reflections of cognition and ideology in science fiction and SF criticism", *Science Fiction Studies*, vol. 6, no. 2, pp. 168–179, 1979.

[ARC 17] ARCHIBUGI D., "Blade runner economics: will innovation lead the economic recovery?", *Research Policy*, vol. 46, no. 3, pp. 535–543, 2017.

[ASI 50] ASIMOV I., *I, Robot*, Doubleday, New York, NY, 1950.

[ATT 05] ATTEBERY B., "Aboriginality in science fiction", *Science Fiction Studies*, vol. 32, no. 3, pp. 385–404, 2005.

[BAN 03] BANERJEE A., "Electricity: science fiction and modernity in early twentieth-century Russia", *Science Fiction Studies*, vol. 30, no. 1, pp. 49–71, 2003.

[BAR 12] BARRETT T.M., "Heart of a serpent? The Cold War science fiction of Murray Leinster", *Science Fiction Studies*, vol. 39, no. 2, pp. 195, 2012.

[BAT 17] BATCHELOR D.A., "Beyond 'Star Trek'", *Communications of the ACM*, vol. 60, no. 5, pp. 104-ff, 2017.

[BEL 95] BELL A., "'Desde Júpiter:' Chile's earliest science-fiction novel", *Science Fiction Studies*, vol. 22, pp. 187–197, 1995.

[BEL 13] BELL F., FLETCHER G., GREENHILL A. *et al.*, "Science fiction prototypes: Visionary technology narratives between futures", *Futures*, vol. 50, pp. 15–24, 2013.

[BER 79] BERGER A.I., "Nuclear energy: science fiction's metaphor of power", *Science Fiction Studies*, vol. 6, no. 2, pp. 121–128, 1979.

[BER 05] BERNE R.W., SCHUMMER J., "Teaching societal and ethical implications of nanotechnology to engineering students through science fiction", *Bulletin of Science, Technology & Society*, vol. 25, no. 6, pp. 459–468, 2005.

[BIE 12] BIELLO D., "It's almost science fiction", *Scientific American*, vol. 306, no. 5, pp. 24, 2012.

[BLA 86] BLACKMORE J.T., LÖWENHARD P., ANDERSON G.L., "Star Wars: science of fiction (with comments and rejoinder)", *International Journal on World Peace*, vol. 3, no. 3, pp. 9–29, 1986.

[BLO 99] BLOOMFIELD B.P., VURDUBAKIS T., "The outer limits: monsters, actor networks and the writing of displacement", *Organization*, vol. 6, no. 4, pp. 625–647, 1999.

[BRO 07] BROADBENT N., "H.G. Wells, Doctor Who and CPR – what do they have in common?", *Manager*, no. 58, pp. 27, 2007.

[BUR 15] BURNAM-FINK M., "Creating narrative scenarios: science fiction prototyping at Emerge", *Futures*, vol. 70, pp. 48–55, 2015.

[BYE 87] BYERS T.B., "Commodity futures: corporate state and personal style in three recent science-fiction movies", *Science Fiction Studies*, vol. 14, no. 3, pp. 326–339, 1987.

[CAS 99] CASE P., "Organizational studies in space: Stanislaw Lem and the writing of social science fiction", *Organization*, vol. 6, no. 4, pp. 649–671, 1999.

[COB 05] COBB M.D., "Framing effects on public opinion about nanotechnology", *Science Communication*, vol. 27, no. 2, pp. 221–239, 2005.

[COL 99] COLATRELLA C., "Science fiction in the information age", *American Literary History*, vol. 11, no. 3, pp. 554–565, 1999.

[COR 95] CORBETT J.M., "Celluloid projections: images of technology and organizational futures in contemporary science fiction film", *Organization*, vol. 2, no. 3/4, pp. 467–488, 1995.

[CRO 10] CROSSAN M.M., APAYDIN M., "A multi-dimensional framework of organizational innovation: a systematic review of the literature", *Journal of Management Studies*, vol. 47, pp. 1154–1191, 2010.

[CSI 03] CSICSERY-RONAY I., "Science fiction and empire", *Science Fiction Studies*, vol. 30, no. 2, pp. 231–245, 2003.

[CSI 96] CSICSERY-RONAY I.J., "The seven beauties of science fiction", *Science Fiction Studies*, vol. 23, no. 3, pp. 385–388, 1996.

[CUS 02] CUSUMANO M.A., GAWER A., *Platform Leadership: How Intel, Microsoft, and Cisco Drive Industry Innovation*, Harvard Business School Press, Brighton, 2002.

[CZA 08] CZARNIAWSKA B., GUSTAVSSON E., "The (d)evolution of the cyberwoman?", *Organization*, vol. 15, no. 5, pp. 665–683, 2008.

[DE 09] DE COCK C., "Jumpstarting the future with Fredric Jameson", *Journal of Organizational Change Management*, vol. 22, no. 4, pp. 437–449, 2009.

[DE 13] DE GRAAF M.M.A., BEN ALLOUCH S., "Exploring influencing variables for the acceptance of social robots", *Robotics and Autonomous Systems*, vol. 61, no. 12, pp. 1476–1486, 2013.

[DEN 01] DENNING P.J., "Many zeros ahead", *Communications of the ACM*, vol. 44, no. 3, p. 112, 2001.

[DEN 04] DENYER D., NEELY A., "Introduction to special issue: innovation and productivity performance in the UK", *International Journal of Management Reviews*, vol. 5–6, pp. 131–135, 2004.

[DIN 15] DINU G., DINU L., "Consumer mind—a real black box", *Ovidius University Annals, Series Economic Sciences*, vol. 15, no. 1, pp. 497–501, 2015.

[DOD 05] DODGSON M., GANN D., Salter A., *Think, Play, Do—Technology, Innovation and Organization*, Oxford University Press, Oxford, 2005.

[DOH 89] DOHERTY T., "Video, science fiction, and the cinema of surveillance", *Journal of the Fantastic in the Arts*, vol. 2, no. 6, pp. 69–79, 1989.

[DON 04] DONAWERTH J., "Body parts: twentieth-century science fiction short stories by women", *PMLA*, vol. 119, no. 3, pp. 474–481, 2004.

[DON 08] DONAWERTH J., "Journal of the fantastic in the arts", *Journal of the Fantastic in the Arts*, vol. 18, no. 2, pp. 152, 2008.

[DOU 12] DOUGHERTY S., "Embodiment and technicity in Geoff Ryman's air", *Science Fiction Studies*, vol. 39, no. 1, pp. 40–59, 2012.

[EDW 14] EDWARDS V.L., "Fifty years of science fiction television", *Administrative Theory & Praxis*, vol. 36, no. 3, pp. 373–397, 2014.

[EMM 97] EMMOTT C., *Narrative Comprehension: A Discourse Perspective*, Clarendon Press, Oxford, 1997.

[ESP 12] ESPOSITO J., "Predicting the present", *Information Services & Use*, vol. 32, no. 1/2, pp. 43–47, 2012.

[ESP 17] ESPY L.D., GUY M.E., "Maybe it is rocket science: how the Martian Reflects Gulick's 'Notes on the Theory of Organization'", *Administrative Theory & Praxis*, vol. 39, no. 3, pp. 193–205, 2017.

[EVA 99] EVANS A.B., "Notable book of SF criticism in France", *Science Fiction Studies*, vol. 26, no. 1, pp. 150–152, 1999.

[FEI 07] FEITELSON D.G., "Asimov's laws of robotics applied to software", *IEEE Software*, vol. 24, no. 4, pp. 112–112, 2007.

[FEK 79] FEKETE J., "'The dispossessed' and 'triton': act and system in utopian science fiction", *Science Fiction Studies*, vol. 6, no. 2, pp. 129–143, 1979.

[FLA 99] FLANAGAN P.S.L., "Cyberspace: the final frontier?", *Journal of Business Ethics*, vol. 19, no. 1, pp. 115–122, 1999.

[FLE 16] FLETCHER G., GREENHILL A., GRIFFITHS M. *et al.*, "Creatively prototyping the future high street", *Production Planning & Control*, vol. 27, no. 6, pp. 477–489, 2016.

[FOS 16] FOSTER R.L., ACKERMAN E., ANDERSON M. *et al.*, "Giving wackiness its due [Back Story]", *IEEE Spectrum*, vol. 53, no. 1, pp. 4, 2016.

[FRA 83] FRANKLIN H.B., "Don't look where we're going: visions of the future in science-fiction films, 1970-82", *Science Fiction Studies*, vol. 10, no. 1, pp. 70–80, 1983.

[FRE 98] FREEDMAN C., "Kubrick's 2001 and the Possibility of a Science-Fiction Cinema", *Science Fiction Studies*, vol. 25, no. 2, pp. 300–318, 1998.

[GEN 17] GENDRON C., IVANAJ S., GIRARD B. *et al.*, "Science-fiction literature as inspiration for social theorizing within sustainability research", *Journal of Cleaner Production*, vol. 164, pp. 1553–1562, 2017.

[GIL 03] GILKS M., ALLEN M., "The subgenres of science fiction", *The Writer*, 2003.

[GOL 01] GOLD J.R., "Under darkened skies: the city in science-fiction film", *Geography*, vol. 86, no. 4, pp. 337–345, 2001.

[GOV 84] GOVAN S.Y., "Connections, links, and extended networks: patterns in Octavia Butler's science fiction", *Black American Literature Forum*, vol. 18, no. 2, pp. 82–87, 1984.

[GRA 15] GRAHAM G., MEHMOOD R., COLES E., "Exploring future cityscapes through urban logistics prototyping: a technical viewpoint", *Supply Chain Management: An International Journal*, vol. 20, no. 3, pp. 341–352, 2015.

[GRA 94] GRAY C.H., "'There will be war!': future war fantasies and militaristic science fiction in the 1980s", *Science Fiction Studies*, vol. 21, no. 3, pp. 315–336, 1994.

[GUT 14] GUTIÉRREZ-JONES C., "Stealing kinship: neuromancer and artificial intelligence", *Science Fiction Studies*, vol. 41, no. 1, pp. 69–92, 2014.

[HAM 03] HAMILTON S.N., "Traces of the future: biotechnology, science fiction, and the media", *Science Fiction Studies*, vol. 30, no. 2, pp. 267–282, 2003.

[HAM 12] HAMNER E., "Remembering the disappeared: science fiction film in postdictatorship Argentina", *Science Fiction Studies*, vol. 39, no. 1, pp. 60, 2012.

[HIG 12] HIGGINS D.M., "Introduction: science fiction and globalization", *Science Fiction Studies*, vol. 39, no. 3, pp. 369–373, 2012.

[HIG 13] HIGGINS D.M., "Psychic decolonization in 1960s science fiction", *Science Fiction Studies*, vol. 40, no. 2, pp. 228, 2013.

[HIG 01] HIGGINS M., "Introduction: more amazing tales", in SMITH W., HIGGINS M., PARKER M. *et al.* (eds), *Science Fiction and Organization*, Routeledge, London, pp. 1–11, 2001.

[HUH 92] HUHTAMO E., *Virtual Zone*, TYY, Turku, 1992.

[HY 76] HY R., "Futures research and policy studies", *Policy Studies Journal*, vol. 4, no. 4, pp. 416–424, 1976.

[IDI 00] IDIER D., "Science fiction and technology scenarios: comparing Asimov's robots and Gibson's cyberspace", *Technology in Society*, vol. 22, no. 2, pp. 255–272, 2000.

[JOH 09] JOHNSON K., "Corporate sperm count and boiled frogs", *International Journal of Contemporary Hospitality Management*, vol. 21, no. 2, pp. 179–190, 2009.

[KAV 01] KAVANAGH D., KEOHANE K., KUHLING C., "Reading star trek. Imagining, theorizing, and reflecting on organizational discourse and practice", in SMITH W., PARKER M., LIGHTFOOT G. (eds), *Science Fiction and Organization*, Routledge, London, pp. 125–142, 2001.

[KEA 99] KEANE J.E., "Piggy and the eternal city: science fiction as testing ground for new management theory", *Emergence*, vol. 1, no. 4, pp. 20–42, 1999.

[KEN 69] KENKEL W.F., "Marriage and the family in modern science fiction", *Journal of Marriage and the Family*, vol. 10, no. 1, pp. 6, 1969.

[KEU 12] KEUPP M.M., PALMIÉ M., GASSMANN O., "The strategic management of innovation: a systematic review and paths for future research", *International Journal of Management Review*, vol. 14, pp. 367–390, 2012.

[KIN 03] KINCAID P., "On the origins of genre", *Extrapolation*, vol. 44, no. 4, pp. 409, 2003.

[KLE 77] KLEIN G., SUVIN D., LECORPS L., "Discontent in American science fiction", *Science Fiction Studies*, vol. 4, no. 1, pp. 3–13, 1977.

[KNI 67] KNIGHT K., "A descriptive model of the intra-firm innovation process", *The Journal of Business*, vol. 40, no. 4, pp. 478–496, 1967.

[KUR 08] KURP P., "Green computing", *Communications of the ACM*, vol. 51, no. 10, pp. 11–13, 2008.

[KUR 05] KURZWEIL R., *The Singularity Is Near: When Humans Transcend Biology*, Penguin, New York, NY, 2005.

[KYM 16] KYMALAINEN T., "Science fiction prototypes as a method for discussing socio-technical issues within emerging technology research and foresight", *Athens Journal of Technology & Engineering*, vol. 3, no. 4, pp. 333–347, 2016.

[LAU 11] LAUDADIO N.C., "'Sounds like a human performance': the electronic music synthesizer in mid-twentieth-century science fiction", *Science Fiction Studies*, vol. 38, no. 2, pp. 304–320, 2011.

[LEF 12] LEFSRUD L.M., MEYER R.E., "Science or science fiction? Professionals' discursive construction of climate change", *Organization Studies*, vol. 33, no. 11, pp. 1477–1506, 2012.

[LEM 85] LEMIEUX J., ROSENTHALL R., "Utopias and social relations in american science fiction, 1950-80", *Science Fiction Studies*, vol. 12, no. 2, pp. 148–155, 1985.

[LOP 04] LOPEZ J., "Bridging the Gaps: Science Fiction in Nanotechnology", *Hyle*, vol. 10, no. 2, pp. 129–152, 2004.

[MA 13] MA S., "'A tale of new Mr. Braggadocio': narrative subjectivity and brain electricity in late qing science fiction", *Science Fiction Studies*, vol. 40, no. 1, pp. 55, 2013.

[MAR 13] MARA M., APPEL M., OGAWA H. *et al.*, "Tell me your story, robot. Introducing an android as fiction character leads to higher perceived usefulness and adoption intention", 2013 *8th ACM/IEEE International Conference on Human-Robot Interaction (HRI)*, pp. 193–194, Tokyo, Japan, 2013.

[MAT 92] MATHESON T.J., "Marcuse, Ellul, and the science-fiction film: negative responses to technology", *Science Fiction Studies*, vol. 19, no. 3, pp. 326–339, 1992.

[MIL 10] MILBURN C., "Modifiable futures: science fiction at the bench", *Isis*, vol. 101, no. 3, pp. 560–569, 2010.

[MOL 69] MOLLOY W.T., "Evaluating the need for automatic data processing equipment in transportation and traffic management", *Transportation Journal*, vol. 8, no. 4, pp. 43–56, 1969.

[MUL 94] MULROW C.D., "Systematic reviews: rationale for systematic reviews", *BMJ*, vol. 309, no. 6954, pp. 597, 1994.

[MUR 07] MURPHY P.C., *What a Book Can Do: The Publication and Reception of Silent Spring*, University of Massachusetts Press, Amherst, 2007.

[MYE 69] MYERS S., MARQUIS D.G., *Successful Industrial Innovation*, National Science Foundation, Washington, 1969.

[NIC 04] NICOLAI A.T., "The bridge to the "real world": applied science or a 'schizophrenic Tour de Force'?", *Journal of Management Studies*, vol. 41, no. 6, pp. 951–976, 2004.

[NIU 08] NIU G.A., "Techno-orientalism, nanotechnology, posthumans, and post-posthumans in Neal Stephenson's and Linda Nagata's science fiction", *MELUS*, vol. 33, no. 4, pp. 73–96, 2008.

[NUD 89] NUDELMAN R., "Soviet science fiction and the ideology of Soviet society", *Science Fiction Studies*, vol. 16, no. 1, pp. 38–66, 1989.

[PAK 16] PAK C., "Terraforming: ecological transformations and environmentalism in science fiction", Liverpool University Press, Liverpool, 2016.

[PAN 89] PANSHIN A., PANSHIN C., *The World beyond the Hill*, Jeremy P. Tarcher, Los Angeles, 1989.

[PAR 08] PARKER M., "Memories of the space age", *Information, Communication & Society*, vol. 11, no. 6, pp. 846–860, 2008.

[PAR 99] PARKER M., HIGGINS M., LIGHTFOOT G. *et al.*, "Amazing tales: organization studies as science fiction", *Organization*, vol. 6, no. 4, pp. 579–590, 1999.

[PEP 17] PEPER E., "Why business leaders need to read more science fiction", *Harvard Business Review*, 2017.

[PHI 99] PHILLIPS N., ZYGLIDOPOULOS S., "Learning from foundation: Asimov's psychohistory and the limits of organization theory", *Organization*, vol. 6, no. 4, pp. 591–608, 1999.

[PIT 04] PITTAWAY L., ROBERTSON M., MUNIR K. *et al.*, "Networking and innovation: a systematic review of the evidence", *International Journal of Management Reviews*, vols 5–6, pp. 137–168, 2004.

[POT 14] POTSTADA M., ZYBURA J., "The role of context in science fiction prototyping: the digital industrial revolution", *Technological Forecasting and Social Change*, vol. 84, pp. 101–114, 2014.

[QUI 85] QUINN J.B., "Managing innovation: controlled chaos", *Harvard Business Review*, vol. 63, no. 3, pp. 73–84, 1985.

[RAB 11] RABINOWITZ P.J., "'The impossible has a way of passing unnoticed': reading science in fiction", *Narrative*, vol. 19, no. 2, pp. 201–215, 2011.

[RAV 17] RAVEN P.G., "Telling tomorrows: science fiction as an energy futures research tool", *Energy Research and Social Science*, vol. 31, pp. 164–169, 2017.

[REE 08] REESE K., "Imagination and realism in Soviet science fiction: Siniavsy's 'Bez Skidok' and Terts's 'Pkhents'", *The Slavic and East European Journal*, vol. 52, no. 3, pp. 439–453, 2008.

[REN 80] RENAULT G., "Science fiction as cognitive estrangement: Darko Suvin and the Marxist critique of mass culture", *Discourse*, vol. 2, pp. 113–141, 1980.

[RIE 10] RIEDER J., "On defining SF, or not: genre theory, SF, and history", *Science Fiction Studies*, vol. 37, no. 2, pp. 191–209, 2010.

[RIV 12] RIVERA L., "Future histories and cyborg labor: reading borderlands science fiction after NAFTA", *Science Fiction Studies*, vol. 39, no. 3, pp. 415, 2012.

[ROB 16a] ROBERTS A., "Definitions", Palgrave Histories of Literature ROBERTS A. (ed.), *The History of Science Fiction,* Palgrave Macmillan, London, pp. 1–23, 2016.

[ROB 16b] ROBERTS A., "From Medieval romance to sixteenth-century Utopia", Palgrave Histories of Literature, ROBERTS A. (ed.), *The History of Science Fiction,* pp. 37–50, Palgrave Macmillan, London, 2016.

[ROG 62] ROGERS E.M., *Diffusion of Innovation*, Free Press, New York, 1962.

[ROT 94] ROTHWELL R., "Towards the Fifth generation innovation process", *International Marketing Review*, vol. 11, no. 1, pp. 7–31, 1994.

[RUS 75] RUSS J., "Towards an aesthetic of science fiction", *Science Fiction Studies*, vol. 2, no. 2, pp. 112–119, 1975.

[RUS 78] RUSS J., "SF and technology as mystification", *Science Fiction Studies*, vol. 5, no. 3, pp. 250–260, 1978.

[SAM 96] SAMUELS D., "'These are the stories that the dogs tell': discourses of identity and difference in ethnography and science fiction", *Cultural Anthropology*, vol. 11, no. 1, pp. 88–118, 1996.

[SAN 77] SANDS S., Warwick K.M., "Successful business innovation", *California Management Review*, vol. 20, no. 2, pp. 5–16, 1977.

[SCH 17] SCHMEINK L., *Biopunk Dystopias : Genetic Engineering, Society and Science Fiction*, Liverpool University Press, Liverpool, 2017.

[SCH 18] SCHNEIDER D., "Ad astra diplomacy [Spectral lines]", *IEEE Spectrum*, vol. 55, no. 1, pp. 6, 2018.

[SCH 99] SCHWARTZMAN R., "Engenderneered machines in science fiction film", *Studies in Popular Culture*, vol. 22, no. 1, pp. 75–87, 1999.

[SEE 10] SEED D., "The course of empire: a survey of the imperial theme in early anglophone science fiction", *Science Fiction Studies*, vol. 37, no. 2, pp. 230–252, 2010.

[SII 96] SIIVOEN T., "Cyborgs and generic oxymorons: the body and technology in William Gibson's cyberspace trilogy", *Science Fiction Studies*, vol. 23, no. 2, pp. 227–244, 1996.

[SIM 09] SIMS C.A., "The dangers of individualism and the human relationship to technology in Philip K. Dick's 'Do Androids Dream of Electric Sheep?'", *Science Fiction Studies*, vol. 36, no. 1, pp. 67–86, 2009.

[SON 13] SONG H., "Chinese science fiction: a response to modernization", *Science Fiction Studies*, vol. 40, no. 1, pp. 15, 2013.

[SRI 99] SRINIVAS N., "Managers as androids: reading moral agency in Philip Dick", *Organization*, vol. 6, no. 4, pp. 609–624, 1999.

[STA 18] STANLEY ROBINSON K., FEDER H., "The realism of our time, Interview with Kim Stanley Robinson", *Radical Philosophy,* vol. 2.01, pp. 87–98, 2018.

[STE 17] STEINMUELLER W.E., "Science fiction and innovation: a response", *Research Policy*, vol. 46, pp. 550–553, 2017.

[STE 14] STEWART C., "Do androids dream of electric free speech? Visions of the future of copyright, privacy and the first amendment in science fiction", *Communication Law and Policy*, vol. 19, no. 4, pp. 433–463, 2014.

[STI 91] STIVALE C.J., "Mille/punks/cyber/plateaus: science fiction and Deleuzo-Guattarian 'becomings'", *SubStance*, vol. 20, no. 3, pp. 66, 1991.

[STO 00] STOCKWELL P., *The Poetics of Science Fiction*, Longman (Pearson Education), London, 2000.

[STO 73] STOVER L.E., "Anthropology and science fiction", *Current Anthropology*, vol. 14, no. 4, pp. 471–474, 1973.

[SUV 81] SUVIN D., "Playful cognizing, or technical errors in Harmonyville: the science fiction of Johanna and Günter Braun", *Science Fiction Studies*, vol. 8, no. 1, pp. 72–79, 1981.

[SUV 82] SUVIN D., "Narrative logic, ideological domination, and the range of science fiction: a hypothesis with a historical test", *Science Fiction Studies*, vol. 9, no. 1, pp. 1–25, 1982.

[TEE 86] TEECE D.J., "Profiting from technological innovation: implications for integration, collaboration, licensing and public policy", *Research Policy*, vol. 15, no. 6, pp. 285–305, 1986.

[TEL 98] TELOTTE J.P., "'So big': the monumental technology of 'things to come'" *Science Fiction Studies*, vol. 25, no. 1, pp. 77–86, 1998.

[THO 05a] THORNE K., "Designing virtual organizations? Themes and trends in political and organizational discourses", *Journal of Management Development*, vol. 24, no. 7, pp. 580–607, 2005.

[THO 05b] THORPE R., HOLT R., MACPHERSON A. *et al.*, "Using knowledge within small and medium-sized firms: a systematic review of the evidence", *International Journal of Management Reviews*, vol. 7, no. 4, pp. 257–281, 2005.

[TIR 99] TIRADO F.J., ALCARAZ J.M., DOMÈNECH M., "A change of episteme for organizations: a lesson from solaris", *Organization*, vol. 6, no. 4, pp. 673–690, 1999.

[TRA 03] TRANFIELD D., DENYER D., SMART P., "Towards a methodology for developing evidence-informed management knowledge by means of systematic review", *British Journal of Management*, vol. 14, no. 3, pp. 207–222, 2003.

[TUR 14] TURNER G., VAN ZOONEN L., HARVEY J., "Confusion, control and comfort: premediating identity management in film and television", *Information, Communication & Society*, vol. 17, no. 8, pp. 986–1000, 2014.

[VAN 03] VAN DER PANNE G., VAN BEERS C., KLEINKNECHT A., "Success and failure of innovation: a literature review", *International Journal of Innovation Management*, vol. 7, no. 3, pp. 309–338, 2003.

[VIN 17] VINGE V., EUCHNER J., "Science fiction as foresight", *Research-Technology Management*, vol. 60, no. 1, pp. 11–17, 2017.

[VOL 13] VOLBERDA H.W., VAN DEN BOSCH F.A.J., HEIJ C.V., "Management innovation: management as fertile ground for innovation", *European Management Review*, vol. 10, pp. 1–15, 2013.

[VON 86] VON HIPPEL E., "Lead users: a source of novel product concepts", *Management Science*, vol. 32, no. 7, pp. 791–805, 1986.

[VOS 85] VOSS C.A., "Determinants of success in the development of applications software", *Journal of Product Innovation Management*, vol. 2, no. 2, pp. 122–129, 1985.

[WEB 08] WEBER R., "A Technology Odyssey", *Journal of Financial Service Professionals*, vol. 62, no. 4, pp. 30–32, 2008.

[WEB 16] WEBER R., "Where no one has gone before: when science fiction inspires technology: EBSCOhost", *Journal of Financial Service Professionals*, vol. 70, no. 4, pp. 43–46, 2016.

[WES 92] WESTFAHL G., "'The Jules Verne, H.G. Wells, and Edgar Allan Poe type of story': Hugo Gernsback's history of science fiction", *Science Fiction Studies*, vol. 19, no. 3, pp. 340–353, 1992.

[WIL 78] WILLIAMS R., "Utopia and science fiction", *Science Fiction Studies*, vol. 5, no. 3, pp. 203–214, 1978.

[WIT 08] WITT M.T., DE HAVEN-SMITH L., "Conjuring the holographic state", *Administration & Society*, vol. 40, no. 6, pp. 547–585, 2008.

[YEO 14] YEOMAN I.S., BEETON S., "The state of tourism futures research", *Journal of Travel Research*, vol. 53, no. 6, pp. 675–679, 2014.

[ZIS 13] ZISER M., "Living with speculative infrastructures", *Boom: A Journal of California*, vol. 3, no. 4, pp. 27–34, 2013.

List of Authors

Sonia ADAM-LEDUNOIS
Equipe MLAB
Laboratoire DRM, CNRS
University Paris-Dauphine
University PSL
France

Florin ALEXA-MORCOV
Faculté de Sociologie et Travail Social
Bucharest
Romania

Marie-Luc ARPIN
Département de management et gestion des ressources humaines
École de gestion de l'Université de Sherbrooke
Canada

Claire AUPLAT
Equipe MLAB
Laboratoire DRM, CNRS
University Paris-Dauphine
University PSL
France

Nadine BOUDOU
LERSEM
Montpellier
France

Marianne CHOUTEAU
S2HEP
INSA Lyon
France

Sébastien DAMART
Equipe MLAB
Laboratoire DRM, CNRS
University Paris-Dauphine
University PSL
France

Corinne GENDRON
Département de stratégie, responsabilité sociale et environnementale
ESG UQÀM
Montreal
Canada

Remy HEMEZ
Armée de Terre
France

Nicolas MERVEILLE
Département de stratégie, responsabilité sociale et environnementale
ESG UQÀM
Montreal
Canada

Thomas MICHAUD
Independent researcher
Poitiers
France

Nicolas MINVIELLE
Audencia Business School
Nantes
France

Céline NGUYEN
S2HEP
INSA Lyon
France

Jean-Pierre REVÉRET
Département de stratégie, responsabilité sociale et environnementale
ESG UQÀM
Montreal
Canada

Olivier WATHELET
Users Matter
Belgium

Index

A, B

C, D

E, F

G, H

I, J

K, M

N, O

P, R

S, T

U, V, Z

Other titles from

ISTE

in

Innovation, Entrepreneurship and Management

2020

ANDREOSSO-O'CALLAGHAN Bernadette, DZEVER Sam, JAUSSAUD Jacques, TAYLOR Richard
Sustainable Development and Energy Transition in Europe and Asia
(Innovation and Technology Set – Volume 9)

CORON Clotilde
Quantifying Human Resources: Uses and Analysis
(Technological Changes and Human Resources Set – Volume 2)

CORON Clotilde, GILBERT Patrick
Technological Change
(Technological Changes and Human Resources Set – Volume 1)

CERDIN Jean-Luc, PERETTI Jean-Marie
The Success of Apprenticeships: Views of Stakeholders on Training and Learning
(Human Resources Management Set – Volume 3)

DIDAY Edwin, GUAN Rong, SAPORTA Gilbert, WANG Huiwen
Advances in Data Science
(Big Data, Artificial Intelligence and Data Analysis Set – Volume 4)

DOS SANTOS PAULINO Victor
Innovation Trends in the Space Industry
(Smart Innovation Set – Volume 25)

GUILHON Bernard
Venture Capital and the Financing of Innovation
(Innovation Between Risk and Reward Set – Volume 6)

MAKRIDES Andreas, KARAGRIGORIOU Alex, SKIADAS Christos H.
Data Analysis and Applications 3: Computational, Classification, Financial, Statistical and Stochastic Methods
(Big Data, Artificial Intelligence and Data Analysis Set – Volume 5)
Data Analysis and Applications 4: Financial Data Analysis and Methods
(Big Data, Artificial Intelligence and Data Analysis Set – Volume 6)

MASSOTTE Pierre, CORSI Patrick
Complex Decision-Making in Economy and Finance

MORLAT Clément
Sustainable Productive System: Eco-development versus Sustainable Development
(Smart Innovation Set – Volume 26)

SAULAIS Pierre, ERMINE Jean-Louis
Knowledge Management in Innovative Companies 2: Understanding and Deploying a KM Plan within a Learning Organization
(Smart Innovation Set – Volume 27)

2019

AMENDOLA Mario, GAFFARD Jean-Luc
Disorder and Public Concern Around Globalization

BARBAROUX Pierre
Disruptive Technology and Defence Innovation Ecosystems
(Innovation in Engineering and Technology Set – Volume 5)

DOU Henri, JUILLET Alain, CLERC Philippe
Strategic Intelligence for the Future 1: A New Strategic and Operational Approach
Strategic Intelligence for the Future 2: A New Information Function Approach

FRIKHA Azza
Measurement in Marketing: Operationalization of Latent Constructs

FRIMOUSSE Soufyane
Innovation and Agility in the Digital Age
(Human Resources Management Set – Volume 2)

GAY Claudine, SZOSTAK Bérangère L.
Innovation and Creativity in SMEs: Challenges, Evolutions and Prospects
(Smart Innovation Set – Volume 21)

GORIA Stéphane, HUMBERT Pierre, ROUSSEL Benoît
Information, Knowledge and Agile Creativity
(Smart Innovation Set – Volume 22)

HELLER David
Investment Decision-making Using Optional Models
(Economic Growth Set – Volume 2)

HELLER David, DE CHADIRAC Sylvain, HALAOUI Lana, JOUVET Camille
The Emergence of Start-ups
(Economic Growth Set – Volume 1)

HÉRAUD Jean-Alain, KERR Fiona, BURGER-HELMCHEN Thierry
Creative Management of Complex Systems
(Smart Innovation Set – Volume 19)

LATOUCHE Pascal
Open Innovation: Corporate Incubator
(Innovation and Technology Set – Volume 7)

LEHMANN Paul-Jacques
The Future of the Euro Currency

LEIGNEL Jean-Louis, MÉNAGER Emmanuel, YABLONSKY Serge
Sustainable Enterprise Performance: A Comprehensive Evaluation Method

LIÈVRE Pascal, AUBRY Monique, GAREL Gilles
Management of Extreme Situations: From Polar Expeditions to Exploration-Oriented Organizations

MILLOT Michel
Embarrassment of Product Choices 2: Towards a Society of Well-being

N'GOALA Gilles, PEZ-PÉRARD Virginie, PRIM-ALLAZ Isabelle
Augmented Customer Strategy: CRM in the Digital Age

NIKOLOVA Blagovesta
The RRI Challenge: Responsibilization in a State of Tension with Market Regulation
(Innovation and Responsibility Set – Volume 3)

PELLEGRIN-BOUCHER Estelle, ROY Pierre
Innovation in the Cultural and Creative Industries
(Innovation and Technology Set – Volume 8)

PRIOLON Joël
Financial Markets for Commodities

QUINIOU Matthieu
Blockchain: The Advent of Disintermediation

RAVIX Joël-Thomas, DESCHAMPS Marc
Innovation and Industrial Policies
(Innovation between Risk and Reward Set – Volume 5)

ROGER Alain, VINOT Didier
Skills Management: New Applications, New Questions
(Human Resources Management Set – Volume 1)

SAULAIS Pierre, ERMINE Jean-Louis
Knowledge Management in Innovative Companies 1: Understanding and Deploying a KM Plan within a Learning Organization
(Smart Innovation Set – Volume 23)

SERVAJEAN-HILST Romaric
Co-innovation Dynamics: The Management of Client-Supplier Interactions for Open Innovation
(Smart Innovation Set – Volume 20)

SKIADAS Christos H., BOZEMAN James R.
Data Analysis and Applications 1: Clustering and Regression, Modeling-estimating, Forecasting and Data Mining
(Big Data, Artificial Intelligence and Data Analysis Set – Volume 2)
Data Analysis and Applications 2: Utilization of Results in Europe and Other Topics
(Big Data, Artificial Intelligence and Data Analysis Set – Volume 3)

VIGEZZI Michel
World Industrialization: Shared Inventions, Competitive Innovations and Social Dynamics
(Smart Innovation Set – Volume 24)

2018

BURKHARDT Kirsten
Private Equity Firms: Their Role in the Formation of Strategic Alliances

CALLENS Stéphane
Creative Globalization
(Smart Innovation Set – Volume 16)

CASADELLA Vanessa
Innovation Systems in Emerging Economies: MINT – Mexico, Indonesia, Nigeria, Turkey
(Smart Innovation Set – Volume 18)

CHOUTEAU Marianne, FOREST Joëlle, NGUYEN Céline
Science, Technology and Innovation Culture
(Innovation in Engineering and Technology Set – Volume 3)

CORLOSQUET-HABART Marine, JANSSEN Jacques
Big Data for Insurance Companies
(Big Data, Artificial Intelligence and Data Analysis Set – Volume 1)

CROS Françoise
Innovation and Society
(Smart Innovation Set – Volume 15)

DEBREF Romain
Environmental Innovation and Ecodesign: Certainties and Controversies
(Smart Innovation Set – Volume 17)

DOMINGUEZ Noémie
SME Internationalization Strategies: Innovation to Conquer New Markets

ERMINE Jean-Louis
Knowledge Management: The Creative Loop
(Innovation and Technology Set – Volume 5)

GILBERT Patrick, BOBADILLA Natalia, GASTALDI Lise,
LE BOULAIRE Martine, LELEBINA Olga
Innovation, Research and Development Management

IBRAHIMI Mohammed
Mergers & Acquisitions: Theory, Strategy, Finance

LEMAÎTRE Denis
Training Engineers for Innovation

LÉVY Aldo, BEN BOUHENI Faten, AMMI Chantal
Financial Management: USGAAP and IFRS Standards
(Innovation and Technology Set – Volume 6)

MILLOT Michel
Embarrassment of Product Choices 1: How to Consume Differently

PANSERA Mario, OWEN Richard
Innovation and Development: The Politics at the Bottom of the Pyramid
(Innovation and Responsibility Set – Volume 2)

RICHEZ Yves
Corporate Talent Detection and Development

SACHETTI Philippe, ZUPPINGER Thibaud
New Technologies and Branding
(Innovation and Technology Set – Volume 4)

SAMIER Henri
Intuition, Creativity, Innovation

TEMPLE Ludovic, COMPAORÉ SAWADOGO Eveline M.F.W.
Innovation Processes in Agro-Ecological Transitions in Developing Countries
(Innovation in Engineering and Technology Set – Volume 2)

UZUNIDIS Dimitri
Collective Innovation Processes: Principles and Practices
(Innovation in Engineering and Technology Set – Volume 4)

VAN HOOREBEKE Delphine
The Management of Living Beings or Emo-management

2017

AÏT-EL-HADJ Smaïl
The Ongoing Technological System
(Smart Innovation Set – Volume 11)

BAUDRY Marc, DUMONT Béatrice
Patents: Prompting or Restricting Innovation?
(Smart Innovation Set – Volume 12)

BÉRARD Céline, TEYSSIER Christine
Risk Management: Lever for SME Development and Stakeholder Value Creation

CHALENÇON Ludivine
Location Strategies and Value Creation of International Mergers and Acquisitions

CHAUVEL Danièle, BORZILLO Stefano
The Innovative Company: An Ill-defined Object
(Innovation between Risk and Reward Set – Volume 1)

CORSI Patrick
Going Past Limits To Growth

D'ANDRIA Aude, GABARRET Inés
Building 21st Century Entrepreneurship
(Innovation and Technology Set – Volume 2)

DAIDJ Nabyla
Cooperation, Coopetition and Innovation
(Innovation and Technology Set – Volume 3)

FERNEZ-WALCH Sandrine
The Multiple Facets of Innovation Project Management
(Innovation between Risk and Reward Set – Volume 4)

FOREST Joëlle
Creative Rationality and Innovation
(Smart Innovation Set – Volume 14)

GUILHON Bernard
Innovation and Production Ecosystems
(Innovation between Risk and Reward Set – Volume 2)

HAMMOUDI Abdelhakim, DAIDJ Nabyla
Game Theory Approach to Managerial Strategies and Value Creation
(Diverse and Global Perspectives on Value Creation Set – Volume 3)

LALLEMENT Rémi
Intellectual Property and Innovation Protection: New Practices and New Policy Issues
(Innovation between Risk and Reward Set – Volume 3)

LAPERCHE Blandine
Enterprise Knowledge Capital
(Smart Innovation Set – Volume 13)

LEBERT Didier, EL YOUNSI Hafida
International Specialization Dynamics
(Smart Innovation Set – Volume 9)

MAESSCHALCK Marc
Reflexive Governance for Research and Innovative Knowledge
(Responsible Research and Innovation Set – Volume 6)

MASSOTTE Pierre
Ethics in Social Networking and Business 1: Theory, Practice and Current Recommendations
Ethics in Social Networking and Business 2: The Future and Changing Paradigms

MASSOTTE Pierre, CORSI Patrick
Smart Decisions in Complex Systems

MEDINA Mercedes, HERRERO Mónica, URGELLÉS Alicia
Current and Emerging Issues in the Audiovisual Industry
(Diverse and Global Perspectives on Value Creation Set – Volume 1)

MICHAUD Thomas
Innovation, Between Science and Science Fiction
(Smart Innovation Set – Volume 10)

PELLÉ Sophie
Business, Innovation and Responsibility
(Responsible Research and Innovation Set – Volume 7)

SAVIGNAC Emmanuelle
The Gamification of Work: The Use of Games in the Workplace

SUGAHARA Satoshi, DAIDJ Nabyla, USHIO Sumitaka
Value Creation in Management Accounting and Strategic Management: An Integrated Approach
(Diverse and Global Perspectives on Value Creation Set –Volume 2)

UZUNIDIS Dimitri, SAULAIS Pierre
Innovation Engines: Entrepreneurs and Enterprises in a Turbulent World
(Innovation in Engineering and Technology Set – Volume 1)

2016

BARBAROUX Pierre, ATTOUR Amel, SCHENK Eric
Knowledge Management and Innovation
(Smart Innovation Set – Volume 6)

BEN BOUHENI Faten, AMMI Chantal, LEVY Aldo
Banking Governance, Performance And Risk-Taking: Conventional Banks Vs Islamic Banks

BOUTILLIER Sophie, CARRÉ Denis, LEVRATTO Nadine
Entrepreneurial Ecosystems (Smart Innovation Set – Volume 2)

BOUTILLIER Sophie, UZUNIDIS Dimitri
The Entrepreneur (Smart Innovation Set – Volume 8)

BOUVARD Patricia, SUZANNE Hervé
Collective Intelligence Development in Business

GALLAUD Delphine, LAPERCHE Blandine
Circular Economy, Industrial Ecology and Short Supply Chains (Smart Innovation Set – Volume 4)

GUERRIER Claudine
Security and Privacy in the Digital Era (Innovation and Technology Set – Volume 1)

MEGHOUAR Hicham
Corporate Takeover Targets

MONINO Jean-Louis, SEDKAOUI Soraya
Big Data, Open Data and Data Development (Smart Innovation Set – Volume 3)

MOREL Laure, LE ROUX Serge
Fab Labs: Innovative User (Smart Innovation Set – Volume 5)

PICARD Fabienne, TANGUY Corinne
Innovations and Techno-ecological Transition (Smart Innovation Set – Volume 7)

2015

CASADELLA Vanessa, LIU Zeting, DIMITRI Uzunidis
Innovation Capabilities and Economic Development in Open Economies (Smart Innovation Set – Volume 1)

CORSI Patrick, MORIN Dominique
Sequencing Apple's DNA

CORSI Patrick, NEAU Erwan
Innovation Capability Maturity Model

FAIVRE-TAVIGNOT Bénédicte
Social Business and Base of the Pyramid

GODÉ Cécile
Team Coordination in Extreme Environments

MAILLARD Pierre
Competitive Quality and Innovation

MASSOTTE Pierre, CORSI Patrick
Operationalizing Sustainability

MASSOTTE Pierre, CORSI Patrick
Sustainability Calling

2014

DUBÉ Jean, LEGROS Diègo
Spatial Econometrics Using Microdata

LESCA Humbert, LESCA Nicolas
Strategic Decisions and Weak Signals

2013

HABART-CORLOSQUET Marine, JANSSEN Jacques, MANCA Raimondo
VaR Methodology for Non-Gaussian Finance

2012

DAL PONT Jean-Pierre
Process Engineering and Industrial Management

MAILLARD Pierre
Competitive Quality Strategies

POMEROL Jean-Charles
Decision-Making and Action

SZYLAR Christian
UCITS Handbook

2011

LESCA Nicolas
Environmental Scanning and Sustainable Development

LESCA Nicolas, LESCA Humbert
Weak Signals for Strategic Intelligence: Anticipation Tool for Managers

MERCIER-LAURENT Eunika
Innovation Ecosystems

2010

SZYLAR Christian
Risk Management under UCITS III/IV

2009

COHEN Corine
Business Intelligence

ZANINETTI Jean-Marc
Sustainable Development in the USA

2008

CORSI Patrick, DULIEU Mike
The Marketing of Technology Intensive Products and Services

DZEVER Sam, JAUSSAUD Jacques, ANDREOSSO Bernadette
Evolving Corporate Structures and Cultures in Asia: Impact of Globalization

2007

AMMI Chantal
Global Consumer Behavior

2006

BOUGHZALA Imed, ERMINE Jean-Louis
Trends in Enterprise Knowledge Management

CORSI Patrick *et al.*
Innovation Engineering: the Power of Intangible Networks

Printed and bound by CPI Group (UK) Ltd, Croydon, CR0 4YY

16/09/2026

14963078-0004